Brooklands Books

DUCATI
Gold Portfolio
1960-1973

Compiled by
R.M.Clarke

ISBN 1 85520 3502

Brooklands Books — BROOKLANDS BOOKS LTD.
P.O. BOX 146, COBHAM,
SURREY, KT11 1LG. UK

Brooklands Books

MOTORING

BROOKLANDS ROAD TEST SERIES

Abarth Gold Portfolio 1950-1971
AC Ace & Aceca 1953-1983
Alfa Romeo Giulietta Gold Portfolio 1954-1965
Alfa Romeo Giulia Berlinas 1962-1976
Alfa Romeo Giulia Coupés 1963-1976
Alfa Romeo Giulia Coupés Gold P. 1963-1976
Alfa Romeo Spider 1966-1990
Alfa Romeo Spider Gold Portfolio 1966-1991
Alfa Romeo Alfasud 1972-1984
Alfa Romeo Alfetta Gold Portfolio 1972-1987
Alfa Romeo Alfetta GTV6 1980-1986
Allard Gold Portfolio 1937-1959
Alvis Gold Portfolio 1919-1967
AMX & Javelin Muscle Portfolio 1968-1974
Armstrong Siddeley Gold Portfolio 1945-1960
Aston Martin Gold Portfolio 1972-1985
Aston Martin Gold Portfolio 1985-1995
Audi Quattro Gold Portfolio 1980-1991
Austin A30 & A35 1951-1962
Austin Healey 100 & 100/6 Gold P. 1952-1959
Austin Healey 3000 Gold Portfolio 1959-1967
Austin Healey Sprite 1958-1971
Barracuda Muscle Portfolio 1964-1974
BMW 1600 Collection No.1 1966-1981
BMW 2002 Gold Portfolio 1968-1976
BMW 316, 318, 320 (4 cyl.) Gold P. 1975-1990
BMW 320, 323, 325 (6 cyl.) Gold P. 1977-1990
BMW M Series Performance Portfolio 1976-1993
BMW 5 Series Gold Portfolio 1981-1987
Bricklin Gold Portfolio 1974-1975
Bristol Cars Gold Portfolio 1946-1992
Buick Automobiles 1947-1960
Buick Muscle Cars 1965-1970
Cadillac Allanté 1986-1993
Cadillac Automobiles 1949-1959
Cadillac Automobiles 1960-1969
Charger Muscle Portfolio 1966-1974
Checker ☆ Limited Edition
Chevrolet 1955-1957
Impala & SS Muscle Portfolio 1958-1972
Chevrolet Corvair 1959-1969
Chevy II & Nova SS Muscle Portfolio 1962-1974
Chevy El Camino & SS 1959-1987
Chevelle & SS Muscle Portfolio 1964-1972
Chevrolet Muscle Cars 1966-1971
Chevy Blazer 1969-1981
Chevrolet Corvette Gold Portfolio 1953-1962
Chevrolet Corvette Sting Ray Gold P. 1963-1967
Chevrolet Corvette Gold Portfolio 1968-1977
High Performance Corvettes 1983-1989
Camaro Muscle Portfolio 1967-1973
Chevrolet Camaro Z28 & SS 1966-1973
Chevrolet Camaro & Z28 1973-1981
High Performance Camaros 1982-1988
Chrysler 300 Gold Portfolio 1955-1970
Chrysler Valiant 1960-1962
Citroen Traction Avant Gold Portfolio 1934-1957
Citroen 2CV Gold Portfolio 1948-1989
Citroen DS & ID 1955-1975
Citroen DS & ID Gold Portfolio 1955-1975
Citroen SM 1970-1975
Cobras & Replicas 1962-1983
Shelby Cobra Gold Portfolio 1962-1969
Cobras & Cobra Replicas Gold P. 1962-1989
Cunningham Automobiles 1951-1955
Daimler SP250 Sports & V-8 250 Saloon Gold P. 1959-1969
Datsun Roadsters 1962-1971
Datsun 240Z 1970-1973
Datsun 280Z & ZX 1975-1983
DeLorean Gold Portfolio 1977-1995
Dodge Muscle Cars 1967-1970
Dodge Viper on the Road
ERA Gold Portfolio 1934-1994
Excalibur Collection No.1 1952-1981
Facel Vega 1954-1964
Ferrari Dino 1965-1974
Ferrari Dino 308 1974-1979
Ferrari 328 • 348 • Mondial Gold Portfolio 1986-1994
Fiat 500 Gold Portfolio 1936-1972
Fiat 600 & 850 Gold Portfolio 1955-1972
Fiat Pininfarina 124 & 2000 Spider 1968-1985
Fiat-Bertone X1/9 1973-1988
Fiat Abarth Performance Portfolio 1972-1987
Ford Consul, Zephyr, Zodiac Mk.I & II 1950-1962
Ford Zephyr, Zodiac, Executive, Mk.III & Mk.IV 1962-1971
Ford Cortina 1600E & GT 1967-1970
High Performance Capris Gold Portfolio 1969-1987
Capri Muscle Portfolio 1974-1987
High Performance Fiestas 1979-1991
High Performance Escorts Mk.I 1968-1974
High Performance Escorts Mk.II 1975-1980
High Performance Escorts 1980-1985
High Performance Escorts 1985-1990
High Performance Sierras & Merkurs Gold Portfolio 1983-1990
Ford Automobiles 1949-1959
Ford Fairlane 1955-1970
Ford Ranchero 1957-1959
Ford Thunderbird 1955-1957
Ford Thunderbird 1958-1963
Ford Thunderbird 1964-1976
Ford GT40 Gold Portfolio 1964-1987
Ford Bronco 1966-1977
Ford Bronco 1978-1988
Goggomobil ☆ Limited Edition
Holden 1948-1962
Honda CRX 1983-1987
International Scout Gold Portfolio 1961-1980
Isetta 1953-1964
ISO & Bizzarrini Gold Portfolio 1962-1974
Jaguar and SS Gold Portfolio 1931-1951
Jaguar XK120, 140, 150 Gold P. 1948-1960
Jaguar Mk.VII, VIII, IX, X, 420 Gold P. 1950-1970
Jaguar Mk.1 & Mk.2 Gold Portfolio 1959-1969
Jaguar C-Type & D-Type ☆ Limited Edition
Jaguar E-Type Gold Portfolio 1961-1971
Jaguar E-Type V-12 1971-1975
Jaguar S-Type & 420 ☆ Limited Edition
Jaguar XJ12, XJ5.3, V12 Gold P. 1972-1990
Jaguar XJ6 Series I & II Gold P. 1968-1979
Jaguar XJ6 Series III 1979-1986
Jaguar XJ6 Gold Portfolio 1986-1994
Jaguar XJS Gold Portfolio 1975-1988
Jaguar XJS Gold Portfolio 1988-1995
Jeep CJ5 & CJ6 1960-1976
Jeep CJ5 & CJ7 1976-1986
Jensen Cars 1946-1967
Jensen Cars 1967-1979
Jensen Interceptor Gold Portfolio 1966-1986
Jensen Healey 1972-1976
Lagonda Gold Portfolio 1919-1964
Lamborghini Countach & Urraco 1974-1980
Lamborghini Countach & Jalpa 1980-1985
Lancia Aurelia & Flaminia Gold Portfolio 1950-1970
Lancia Fulvia Gold Portfolio 1963-1976
Lancia Beta Gold Portfolio 1972-1984
Lancia Delta Gold Portfolio 1979-1994
Lancia Stratos 1972-1985
Land Rover Series I 1948-1958
Land Rover Series II & IIa 1958-1971
Land Rover Series III 1971-1985
Land Rover 90 110 Defender Gold Portfolio 1983-1994
Land Rover Discovery 1989-1994
Land Rover Story Part One 1948-1971
Lincoln Gold Portfolio 1949-1960
Lincoln Continental 1961-1969
Lincoln Continental 1969-1976
Lotus Sports Racers Gold Portfolio 1953-1965
Lotus Seven Gold Portfolio 1957-1974
Lotus Caterham Seven Gold Portfolio 1974-1995
Lotus Elite & Eclat 1974-1982
Lotus Elan Gold Portfolio 1962-1974
Lotus Elan Collection No. 2 1963-1972
Lotus Elan & SE 1989-1992
Lotus Cortina Gold Portfolio 1963-1970
Lotus Europa Gold Portfolio 1966-1975
Lotus Elite & Eclat 1974-1982
Lotus Turbo Esprit 1980-1986
Marcos Cars 1960-1988
Maserati 1965-1970
Mazda RX-7 Gold Portfolio 1978-1991
Mercedes 190 & 300 SL 1954-1963
Mercedes 230/250/280SL 1963-1971
Mercedes G Wagen 1981-1994
Mercedes Benz SLs & SLCs Gold P. 1971-1989
Mercedes S & 600 1965-1972
Mercedes S Class 1972-1979
Mercedes SLs Performance Portfolio 1989-1994
Mercury Muscle Cars 1966-1971
Messerschmitt Gold Portfolio 1954-1964
MG Gold Portfolio 1929-1939
MG TA & TC Gold Portfolio 1936-1949
MG TD & TF Gold Portfolio 1949-1955
MGA & Twin Cam Gold Portfolio 1955-1962
MG Midget Gold Portfolio 1961-1979
MGB Roadsters 1962-1980
MGB MGC & V8 Gold Portfolio 1962-1980
MGB GT 1965-1980
MG Y-Type & Magnette ZA/ZB ☆ Limited Edition
Mini Gold Portfolio 1959-1969
Mini Gold Portfolio 1969-1980
High Performance Minis Gold Portfolio 1960-1973
Mini Cooper Gold Portfolio 1961-1971
Mini Moke Gold Portfolio 1964-1994
Mopar Muscle Cars 1964-1967
Morgan Three-Wheeler Gold Portfolio 1910-1952
Morgan Plus 4 & Four 4 Gold P. 1936-1967
Morgan Cars 1960-1970
Morgan Cars Gold Portfolio 1968-1989
Morris Minor Collection No. 1 1948-1980
Shelby Mustang Muscle Portfolio 1965-1970
High Performance Mustang IIs 1974-1978
High Performance Mustangs 1982-1988
Nash-Austin Metropolitan Gold P. 1954-1962
Oldsmobile Automobiles 1955-1963
Oldsmobile Muscle Cars 1964-1971
Oldsmobile Toronado 1966-1978
Opel GT Gold Portfolio 1968-1973
Packard Gold Portfolio 1946-1958
Pantera Gold Portfolio 1970-1989
Panther Gold Portfolio 1972-1990
Plymouth Muscle Cars 1966-1971
Pontiac Tempest & GTO 1961-1965
Pontiac Muscle Cars 1966-1972
Pontiac Firebird & Trans-Am 1973-1981
High Performance Firebirds 1982-1988
Pontiac Fiero 1984-1988
Porsche 356 Gold Portfolio 1953-1965
Porsche 911 1965-1969
Porsche 911 1970-1972
Porsche 911 1973-1977
Porsche 911 Turbo 1975-1984
Porsche 911 SC & Turbo Gold Portfolio 1978-1983
Porsche 911 Carrera & Turbo Gold P. 1984-1989
Porsche 924 Gold Portfolio 1975-1988
Porsche 928 Performance Portfolio 1977-1994
Porsche 944 Gold Portfolio 1981-1991
Range Rover Gold Portfolio 1970-1985
Range Rover Gold Portfolio 1986-1995
Reliant Scimitar 1964-1986
Riley Gold Portfolio 1924-1939
Riley 1.5 & 2.5 Litre Gold Portfolio 1945-1955
Rolls Royce Silver Cloud & Bentley 'S' Series Gold Portfolio 1955-1965
Rolls Royce Silver Shadow Gold P. 1965-1980
Rolls Royce & Bentley Gold P. 1980-1989
Rover P4 1949-1959
Rover P4 1955-1964
Rover 3 & 3.5 Litre Gold Portfolio 1958-1973
Rover 2000 & 2200 1963-1977
Rover 3500 1968-1977
Rover 3500 & Vitesse 1976-1986
Saab Sonett Collection No.1 1966-1974
Saab Turbo 1976-1983
Studebaker Gold Portfolio 1947-1966
Studebaker Hawks & Larks 1956-1963
Avanti 1962-1990
Sunbeam Tiger & Alpine Gold P. 1959-1967
Toyota MR2 1984-1988
Toyota Land Cruiser 1956-1984
Triumph Dolomite Sprint ☆ Limited Edition
Triumph TR2 & TR3 Gold Portfolio 1952-1961
Triumph TR4, TR5, TR250 1961-1968
Triumph TR6 Gold Portfolio 1969-1976
Triumph TR7 & TR8 Gold Portfolio 1975-1982
Triumph Herald 1959-1971
Triumph Vitesse 1962-1971
Triumph Spitfire Gold Portfolio 1962-1980
Triumph 2000, 2.5, 2500 1963-1977
Triumph GT6 Gold Portfolio 1966-1974
Triumph Stag Gold Portfolio 1970-1977
TVR Gold Portfolio 1959-1986
TVR Performance Portfolio 1986-1994
VW Beetle Gold Portfolio 1935-1967
VW Beetle Gold Portfolio 1968-1991
VW Beetle Collection No.1 1970-1982
VW Karmann Ghia 1955-1982
VW Bus, Camper, Van 1954-1967
VW Bus, Camper, Van 1968-1979
VW Bus, Camper, Van 1979-1989
VW Scirocco 1974-1981
VW Golf GTI 1976-1986
Volvo PV444 & PV544 1945-1965
Volvo Amazon-120 Gold Portfolio 1956-1970
Volvo 1800 Gold Portfolio 1960-1973
Volvo 140 & 160 Series Gold Portfolio 1966-1975

Forty Years of Selling Volvo

BROOKLANDS ROAD & TRACK SERIES

Road & Track on Alfa Romeo 1964-1970
Road & Track on Alfa Romeo 1971-1976
Road & Track on Aston Martin 1962-1990
R & T on Auburn Cord and Duesenburg 1952-84
Road & Track on Audi & Auto Union 1952-1980
Road & Track on Audi & Auto Union 1980-1986
Road & Track on Austin Healey 1953-1970
Road & Track on BMW Cars 1966-1974
Road & Track on BMW Cars 1975-1978
Road & Track on BMW Cars 1979-1983
R & T on Cobra, Shelby & Ford GT40 1962-1992
Road & Track on Corvette 1953-1967
Road & Track on Corvette 1968-1982
Road & Track on Corvette 1982-1986
Road & Track on Corvette 1986-1990
Road & Track on Ferrari 1975-1981
Road & Track on Ferrari 1981-1984
Road & Track on Ferrari 1984-1988
Road & Track on Fiat Sports Cars 1968-1987
Road & Track on Jaguar 1950-1960
Road & Track on Jaguar 1961-1968
Road & Track on Jaguar 1968-1974
Road & Track on Jaguar 1974-1982
Road & Track on Jaguar 1983-1989
Road & Track on Lamborghini 1964-1985
Road & Track on Lotus 1972-1981
Road & Track on Maserati 1975-1983
R & T on Mazda RX7 & MX5 Miata 1986-1991
Road & Track on Mercedes 1952-1962
Road & Track on Mercedes 1963-1970
Road & Track on Mercedes 1971-1979
Road & Track on Mercedes 1980-1987
Road & Track on MG Sports Cars 1949-1961
Road & Track on MG Sports Cars 1962-1980
Road & Track on Mustang 1964-1977
R & T on Nissan 300-ZX & Turbo 1984-1989
Road & Track on Pontiac 1960-1983
Road & Track on Porsche 1951-1967
Road & Track on Porsche 1968-1971
Road & Track on Porsche 1972-1975
Road & Track on Porsche 1975-1978
Road & Track on Porsche 1985-1988
R & T on Rolls Royce & Bentley 1950-1965
R & T on Rolls Royce & Bentley 1966-1984
Road & Track on Saab 1972-1992
R & T on Toyota Sports & GT Cars 1966-1984
R & T on Triumph Sports Cars 1953-1967
R & T on Triumph Sports Cars 1967-1974
R & T on Triumph Sports Cars 1974-1982
Road & Track on Volkswagen 1951-1968
Road & Track on Volkswagen 1968-1978
Road & Track on Volkswagen 1978-1985
Road & Track on Volvo 1957-1974
Road & Track on Volvo 1977-1994
R&T - Henry Manney at Large & Abroad
R&T - Peter Egan's "Side Glances"
R&T - Peter Egan "At Large"

BROOKLANDS CAR AND DRIVER SERIES

Car and Driver on BMW 1955-1977
Car and Driver on BMW 1977-1985
C and D on Cobra, Shelby & Ford GT40 1963-84
Car and Driver on Corvette 1978-1982
Car and Driver on Corvette 1983-1988
C and D on Datsun Z 1600 & 2000 1966-1984
Car and Driver on Ferrari 1955-1962
Car and Driver on Ferrari 1963-1975
Car and Driver on Ferrari 1976-1983
Car and Driver on Mopar 1956-1967
Car and Driver on Mopar 1968-1975
Car and Driver on Mustang 1964-1972
Car and Driver on Pontiac 1961-1975
Car and Driver on Porsche 1955-1962
Car and Driver on Porsche 1963-1970
Car and Driver on Porsche 1970-1976
Car and Driver on Porsche 1977-1981
Car and Driver on Porsche 1982-1986
Car and Driver on Saab 1956-1985
Car and Driver on Volvo 1955-1986

BROOKLANDS PRACTICAL CLASSICS SERIES

PC on Austin A40 Restoration
PC on Land Rover Restoration
PC on Metalworking in Restoration
PC on Midget/Sprite Restoration
PC on MGB Restoration
PC on Sunbeam Rapier Restoration
PC on Triumph Herald/Vitesse
PC on Spitfire Restoration
PC on Beetle Restoration
PC on 1930s Car Restoration

BROOKLANDS HOT ROD 'MUSCLECAR & HI-PO ENGINES' SERIES

Chevy 265 & 283
Chevy 302 & 327
Chevy 348 & 409
Chevy 350 & 400
Chevy 396 & 427
Chevy 454 thru 512
Chrysler Hemi
Chrysler 273, 318, 340 & 360
Chrysler 361, 383, 400, 413, 426, 440
Ford 289, 302, Boss 302 & 351W
Ford 351C & Boss 351
Ford Big Block

BROOKLANDS RESTORATION SERIES

Auto Restoration Tips & Techniques
Basic Bodywork Tips & Techniques
Camaro Restoration Tips & Techniques
Chevrolet High Performance Tips & Techniques
Chevy Engine Swapping Tips & Techniques
Chevy-GMC Pickup Repair
Chrysler Engine Swapping Tips & Techniques
Engine Swapping Tips & Techniques
Ford Pickup Repair
How to Build a Street Rod
Land Rover Restoration Tips & Techniques
MG 'T' Series Restoration Guide
MGA Restoration Guide
Mustang Restoration Tips & Techniques
Performance Tuning - Chevrolets of the '60's
Performance Tuning - Pontiacs of the '60's

MOTORCYCLING

BROOKLANDS ROAD TEST SERIES

AJS & Matchless Gold Portfolio 1945-1966
BSA Twins A7 & A10 Gold Portfolio 1946-1962
BSA Twins A50 & A65 Gold Portfolio 1962-1973
Ducati Gold Portfolio 1960-1974
Ducati Gold Portfolio 1974-1978
Laverda Gold Portfolio 1967-1977
Norton Commando Gold Portfolio 1968-1977
Triumph Bonneville Gold Portfolio 1959-1983

BROOKLANDS CYCLE WORLD SERIES

Cycle World on BMW 1974-1980
Cycle World on BMW 1981-1986
Cycle World on Ducati 1982-1991
Cycle World on Harley-Davidson 1962-1968
Cycle World on Harley-Davidson 19781-1983
Cycle World on Harley-Davidson 1983-1987
Cycle World on Harley-Davidson 1987-1990
Cycle World on Harley-Davidson 1990-1992
Cycle World on Honda 1962-1967
Cycle World on Honda 1968-1971
Cycle World on Honda 1971-1974
Cycle World on Husqvarna 1966-1976
Cycle World on Husqvarna 1977-1984
Cycle World on Kawasaki 1966-1971
Cycle World on Kawasaki Off-Road Bikes 1972-1979
Cycle World on Kawasaki Street Bikes 1972-1976
Cycle World on Norton 1962-1971
Cycle World on Suzuki 1962-1970
Cycle World on Suzuki Off-Road Bikes 1971-1976
Cycle World on Suzuki Street Bikes 1971-1976
Cycle World on Triumph 1967-1972
Cycle World on Yamaha 1962-1969
Cycle World on Yamaha Off-Road Bikes 1970-1974
Cycle World on Yamaha Street Bikes 1970-1974

MILITARY

BROOKLANDS MILITARY VEHICLES SERIES

Allied Military Vehicles No.2 1941-1946
Complete WW2 Military Jeep Manual
Dodge Military Vehicles No.1 1940-1945
Hail To The Jeep
Land Rovers in Military Service
Military & Civilian Amphibians 1940-1990
Off Road Jeeps: Civ. & Mil. 1944-1971
US Military Vehicles 1941-1945
US Army Military Vehicles WW2-TM9-2800
VW Kubelwagen Military Portfolio 1940-1990
WW2 Jeep Military Portfolio 1941-1945

CONTENTS

Page	Article	Publication	Date		
5	The Ducati 'Super Sports'	Motor Cycling	Mar	3	1960
7	Watch this Strip down of a Ducati	Motorcycle Mechanics	July		1962
10	249cc Ducati Daytona Road Test	Motor Cycling	Sept	7	1961
12	Ducati Scrambler 250cc Road Test	Cycle World	Aug		1962
16	Ducati Daytona 250cc Road Test	Motorcycle Mechanics	Sept		1962
18	Ducati 250 Monza Road Test	Cycle World	May		1963
22	Ducati 48cc Sports	Motor Cycle	June	27	1963
23	Trailering with the Ducati Bronco Tour Test	Cycle	Oct		1963
26	Ducati 50 Road Test	Motorcycle Mechanics	Oct		1963
28	Ducati Diana Mk 3 Road Test	Cycle World	Dec		1963
32	Testing the Ducati "Mountaineer"	Cycle	Jan		1964
35	New Ducati Four	Cycle World	April		1964
36	249cc Ducati Daytona Road Test	Motor Cycle	May	14	1964
38	249cc Ducati Mach 1 Road Test	Motor Cycle	Nov	5	1964
40	Ducati Diana Mk III Road Test	Cycle World	Nov		1964
44	Ducati 200 SS Road Test	Motorcycle Mechanics	Jan		1965
46	Ducati Cadet 90	Cycle World	Mar		1965
48	New Ducati Models	Cycle	April		1965
49	Ducati Portable Scooter Test	Cycle World	Mar		1965
50	Ducati 160 Monza Jr. Road Test	Cycle World	July		1965
53	Doing a Ducati Strip down of 200 & 250cc	Motorcycle Mechanics	Oct		1965
56	Ducati 250 Mk III Road Test	Cycle World	Aug		1965
60	Ducati Daytona 250cc Road Test	Motorcycle Mechanics	Nov		1965
62	A Ducati 250 for Racing Part 1	Cycle World	Dec		1965
64	New Ducati Models	Cycle	Feb		1966
65	Ducati 250cc 5-Speed Scrambler Road Test	Cycle	Jan		1966
68	A Ducati 250 for Racing Part 2	Cycle World	Feb		1966
72	A Ducati 250 for Racing Part 3	Cycle World	Oct		1966
74	Customized Ducati 200 Super Sports Road Test	Motor Cycle	Aug	25	1966
77	Ducati 160 Monza Junior	Motor Cyclist Illustrated	Mar		1967
78	Ducati 160 Monza Junior Road Test	Cycle	Jan		1968
82	Well, it's one way to Spend a Rainy week in November 350 Ducati Sebring Road Test	Motorcycle Sport	Jan		1968
88	Dashing Ducati 350cc Track Test	Motorcycle Mechanics	June		1968
90	Ducati Fork Strip	Motorcycle Mechanics	April		1969
92	Ducati 250s Sportster vs Roadsters Monza vs Mark 3 Comparison Test	Motorcycle Mechanics	Sept		1969
95	Ducati 350 SSS Road Test	Cycle	Feb		1969
100	Ray Knight Race-Tests the Desmodromic Ducati	Motor Cyclist Illustrated	Nov		1969
103	Three Desmo Ducatis: The Mark 3D 250, 350, 450 Road Test	Cycle	Feb		1970
109	Ducati 250 & 350 Engine Analysis	Motorcycle Mechanics	Aug		1970
113	Desmo Ducati Engine Dismantling	Motorcycle Mechanics	Feb		1970
116	The Ducati that Ron Built - 391cc Dirt Special	Cycle World	Dec		1970
118	Ducati 450 R/T Road Test	Cycle World	Sept		1971
122	Ducati 250 24 Hours	Motorcycle Sport	April		1972
124	Satisfied Mind	Cycle	July		1972
132	Long Road to Imola	Cycle	Aug		1972
139	One Breathes as well as Two... Racing a Ducati 250	Motorcycle Sport	June		1972
141	Preview: The Ducati 750	Cycle World	Sept		1972
146	Ducati 750 V-Twin Road Test	Cycle	Oct		1972
152	One for the Road Ducati 750 Sport	Cycle	Jan		1973
155	Mick Walker's 250 Ducati	Motorcycle Sport	July		1973
159	New Tricks from Italy - Ducati's 1973 Desmo Racer	Cycle	Aug		1973
163	Ducati	Cycle Buyers Guide			1973
164	Ducati History	Cycle World	Dec		1966
170	Long Playing Single	Motorcycle Mechanics	May		1979

ACKNOWLEDGEMENTS

Brooklands Books have long been recognised in the automotive world as a valuable source of historical information on the finest and most sought after cars. With more than 30 years' experience in compiling these reference anthologies motorcyclists thought it was time that they turned their attention to motorcycles and motorcycling, and requested Brooklands to do for bikes, what they had done for cars. Our first books in this series covered the post-war classic British machines and now we turn our attention to famous Italian marques.

We have invited Jeff Clew, a leading authority on vintage and classic motorcycles, to write this introduction on the early Ducati models. His specialised knowledge of these machines is not limited to his reputation as an historian and restorer, as he is currently enjoying his 50th consecutive year of riding motorcycles of all kinds, both on and off-road.

Like the other titles in the Brooklands Books library, this latest addition has been compiled with the help and generous co-operation of the world's leading motorcycle magazine publishers. We are indebted in this instance to the managements of *Cycle, Cycle Buyer's Guide, Cycle World, Motor Cycle, Motor Cycling, Motorcycle Mechanics, Motorcycle Sport* and *Motor Cyclist Illustrated* for permission to include their copyright material.

Our thanks also go to David Harvey of the Ducati Owners Club GB, who at short notice, kindly supplied us with the excellent photographs which illustrate our front and back covers.

R.M. Clarke

It was not until 1950 that Ducati Meccanica S.p.A. of Bologna made their first appearance on the motorcycle scene with a 60cc motorcycle. Until then, their main occupation had been the manufacture of electrical equipment, which extended back into the pre-World War II days. Their first offering to the U.K. market was a 48cc cyclemotor engine unit, which was attached to a conventional bicycle and sold by Britax (London) Ltd., who acted as their concessionaires. Even that tiny power unit showed some of Ducati's ingenuity, for it was an ohv design built in-unit with a two-speed gear. Many still regard it as the best of the cyclemotors.

As the cyclemotor progressed to the fully integrated moped, Ducati were soon able to offer comparable models and as early as 1957 had 124cc and 174cc motorcycles powered by an overhead camshaft engine, with a duplex tube frame, telescopic front forks, four-speed gearbox and pivoted fork rear suspension. Such was their quality and styling that they soon gave the British motorcycle industry cause for concern.

A whole rash of models appeared in the years that followed, so many that the provision of spare parts for them was unable to keep pace. Fortunately, these problems were eventually overcome and it was possible to obtain models of up to 450cc capacity. Needless to say it was the overhead camshaft models that appealed to the more discerning riders, who were well aware of the company's competition successes in road racing events of international status.

This success could be largely attributed to Ing. Fabio Taglioni, who joined Ducati in 1954. Ducati played a leading role in the development of desmodromic valve gear, in which the valves are opened and closed by separate cams. This ensures more accurate valve timing at high rpm, without the usual limitations of valve bounce and other factors. Even the standard production overhead camshaft engines had a camshaft assembly driven by spiral bevels and a vertical shaft, despite the need for individual fits and hand assembly. Ducati were never a company to spoil a promising design by compromise or by sacrificing quality.

Later, the company progressed to produce some very memorable large capacity vee twins. They, however, form the main subject of a second Ducati book that runs parallel to this and continues the Ducati story from 1974 onwards.

Jeff Clew

Cover photographs all supplied and taken by David Harvey. Front cover: Steve Hope riding his 1973 450 single. Back cover - Top: Bernard Lambs 250 single (circa 1968). Back cover - Bottom: Andy Jacobsons 1972 750GT.

IMPRESSIONS OF 1960 MODELS

The DUCATI 'SUPER SPORTS'

A fast 200 c.c. o.h.c. Italian roadster

IN appearance the "Super Sports" Ducati epitomizes the traditional Italian lightweight roadburner — big, recessed tank, crouched riding position, bronze and flamboyant red *décor*. To the conservative British eye, the total effect may be ostentatious; but riding experience soon removes any suspicion that this is merely a slice of Latin café-racer's delight.

The "Super Sports" is a really efficient tool for fast road work—a 200 c.c. o.h.c. device capable of reaching the upper sixties with full road-going equipment and silencing—and with acceleration and road holding to match. Braking, too, is definitely in the race-bred category.

The test machine was kindly lent to *Motor Cycling* by its owner, Mr. E. Smith of London, E.14.

This Ducati is obviously a one-man mount. It is true that a full-length dual seat is fitted, as well as pillion footrests. But these, plainly, are meant primarily to complement the dropped handlebars and to assist the adoption of a racing crouch. Accommodation for a passenger is a secondary, and very minor, consideration. However, it can be provided—with a little discomfort to both parties.

The certified kerbside weight with one gallon of fuel and oil aboard was 273 lb., a figure indicative of intelligent weight-saving on the part of the designer. With the rider seated realistically rearward, the front wheel carried 38% of the total weight, the rear 62%. This apparently inequitable distribution did not hamper roadholding or steering.

Throwing It About

One's first impression was that tautness and accuracy of steering were *not* on the Ducati's list of credit points. But this sensation lasted for only 50 miles or so, by which time the tester had gained the "feel" of his mount. Then the sense of instability waned and confidence grew with every passing mile.

The fact is that the "Super Sports" needs to be steered. Once this had been grasped, and the knack of using the handlebars as a rudder-bar as well as a banking lever mastered, the Ducati was really thrown about. And how it revelled in it!

Corners could be taken fast. And we mean fast. Plenty of ground clearance permitted the sort of lean that scuffs the outside of the welt off the shoes. And if lock was fed on with the lean, stability was assured. The shaped, 3¾ gal. tank did not foul the thumbs when the adjustable bars were on full lock.

On wet surfaces, handling was equally good—with the obvious qualification, of course! In the second 100 miles of transient ownership, the tester was taking the Ducati up to wheel-breakaway point. It was not at all hard to feel this approaching; the sensation was one of both wheels wanting

The 180-mm. front brake is equipped with cast-in air intake and exhaust vents.

The brake-plate of the rear full-width unit is on the left, the sprocket on the right; the all-indirect gearbox is of the crossover type. (Right) Clutch adjustment cover removed; Allen Key on the plug which shrouds an extractor thread for the primary-gear cover.

to "go" together if the rider were foolhardy enough to increase his rate of turn past the limit of tyre adhesion.

Braking on rain-sodden roads was also first-class. In fact, stopping ability in the rain, one felt, came close to setting new test standards. The front anchor and tyre seemed to have a welcome abundance of grip.

The front brake is ventilated by a cast-in air scoop. Thanks to this, and the general adequacy of brake dimensions, there was no fade under hard-riding conditions. The back brake played its part, being sweet and controllable. Dip of the "teles." under hard braking was so unobtrusive that one had to look for it deliberately.

The anchors were perfect from speed. It was a pleasant task to lop, say, 20 m.p.h. from some 70 m.p.h. when descending a hill with a curve at the bottom. No judder was experienced.

However, a species of judder unconnected with braking did set in over a certain type of surface. This took the form of rapid jarring up and down of the front forks when passing over ripples about 12 to 18 inches apart at some 25 m.p.h. It was decidedly uncomfortable but never threatened control.

The rear springing was firm. Some welcome comfort was imparted by the built-in springs of the dual seat. The rider was also able to obtain a firm grip of the tank via the knee recesses. The position of the bars and the rests generally encouraged a rider-wedged-in-and-part-of-the-machine feel.

As the "Super Sports" was found to be capable of notching a two-way-mean top speed of 66 m.p.h., this high degree of rider-attachment was, of course, a Good Thing. It was important, too, when using the excellent acceleration available. The standing quarter-mile could be covered in 22.0 seconds, 56 m.p.h. being recorded at the end.

The 100 m.p.h. speedometer showed 35 m.p.h. at a true 30, and 76 m.p.h. at a

true 63, the latter being the figure most frequently rung up on the open road after a burst of acceleration.

Gear ratios were ideal. Bottom was not too low for an urgent departure, nor too high for a get-away in traffic on a steep adverse gradient. A true 44 m.p.h. was usually notched in second, and 59 in third. Optimum use of engine output demanded a readiness to "play tunes" on the two closely spaced upper ratios.

Cog-swapping between any pair of gears could only be described as an enthusiasts' delight. The clutch was normally dispensed with on the upward changes from second and from third. The gears then slid in as sleekly as on any machine in the tester's experience.

Regrettably, the internal mechanism was lazy in its self-centring action after a downward change; the external rocking pedal did return, however. The toe portion of this component, depressed for higher ratios, was beautifully positioned but the heel part was awkwardly sited, so lower ratios also were selected with the toe.

Clutch action was perfection itself. Demanding a light-to-medium operating pressure, it disengaged instantly and cleanly on the first kick of the morning.

Cold engine starting drill was orthodox and swift; hot starting was equally facile. The knack of using a left-hand-sited kick-starter was soon mastered. But the trick of getting the Ducati on to its high-lift centre stand was not. Once on the stand, there was firm support.

Lighting was satisfactory. Particularly efficient were the shapes of the two beams—a clean, wide rectangle on the main beam and a flat-topped semi-circle on dip. The horn, too, did its job well.

Fuel consumption ran out at a creditable 59 m.p.g. for the type of riding in which this sporting mount would usually be engaged. Some credit for this must go to the flexible nature of the engine, which permitted top gear to be used as a practical ratio for traffic-crawling down to 24 m.p.h. As idling was reliable, there was no need for throttle-blipping—often a profligate consumer of fuel.

Though vibration never dictated the speed to be set, exhaust noise frequently did. Neither was the engine mechanically silent —hardly surprising for a sports motor with a high-compression large-clearance piston and "hot" valve gear. The valve springs, apparently, are strong enough to prevent valve bounce or float, the ceiling revs. in the gears being set by the engine breathing.

To the true owner of 272 JKK, our envious thanks. We can well understand the pleasure that proprietorship of this mount must give him, and are most grateful for the opportunity of sampling once more the satisfaction that comes from riding one of these unburstable, ever-revving Ducatis.

SPECIFICATION

ENGINE
Type	Single-cylinder four-stroke
Bore	67 mm.
Stroke	57.8 mm.
Cubic capacity	204 c.c.
Valves	Overhead with o.h.c.
Compression ratio	8.5 : 1
Carburetter	Dell 'Orto UBF24BS
Ignition	Coil, automatic control
Maker's claimed output	17 b.h.p./7,500 r.p.m.
Lubrication	Wet sump, gear pump
Starting	Kick-starter

TRANSMISSION
Unit construction gearbox with footchange
Ratios	8.5, 10.5, 14.7, 24.5 : 1
Speed at 1,000 r.p.m. in top gear	8 m.p.h.
Primary drive	Helical-tooth spur gears
Final drive	Chain
Clutch	Multi-plate in oilbath
Shock-absorber	Cush-drive rear hub

CYCLE PARTS
Frame	Tubular backbone type, single front down-tube, engine unit forming bottom member
Front suspension	Telescopic; coil springs, hydraulically damped on both strokes
Rear suspension	Swinging-fork; hydraulically damped by Marzocchi units
Wheelbase	53 in.
Tyres	Pirelli interrupted-ribbed front, 2.50 × 18 in.; studded rear, 2.75 × 18 in.
Brakes	Front, 7½ in. dia.; rear, 6½ in. dia.; full-width hubs
Fuel tank	Single tap; two-bolt and spring-strap fixing; rubber mounted
Oil sump	3½ pints
Generator	Cev 6-v. A.C. generator with rectifier for D.C. battery lighting
Lamps	20/20-w. head, 3-w. pilot, 10/3-w. stop-tail
Battery	Smiths 13 a.h.
Speedometer	Veglia, 100 m.p.h. (non-trip)
Seating	Single-level dual seat
Stands	Centre
Standard finish	Bronze and flamboyant red
Certified kerbside weight	273 lb.

EQUIPMENT
Dropped bars; pillion rests; toe-heel gear pedal; transverse front registration plate

PRICES
Machine	£274 4s. 7d. (inc. £46 17s. 9d. P.T.)
Extras	None
Tax	£1 17s. 6d. p.a.
Makers	Ducati Meccanica S.p.A., Bologna, Italy

Concessionnaires: Ducati Concessionaires, Ltd., 80 Burleigh Road, Stretford, Lancs.

photo guide

WATCH THIS STRIPDOWN OF A DUCATI

DAVID WEIGHTMAN DESCRIBES THE TWENTY-FOUR STAGES

DUCATI is a name which has long been associated with high performance motorcycles and the 250 c.c. Daytona is one of the most popular. This overhead camshaft engined machine is capable of about 90 m.p.h. and a racing kit is available!

Getting the engine out of the frame is a simple job and can be done quite quickly by removing the following parts—petrol tank, dual seat, rev. counter drive, exhaust pipe, coil, carburetter, rear chain, and disconnecting all wires to the engine. In particular the two wires to the rectifier must be removed—they are both yellow and are interchangeable.

Assuming that the engine has already been cleaned down and the oil drained off, it can now be lifted out and put on the bench for the strip proper. Some special tools are required but these can be bought or hired from your dealer or Ducati Concessionaires.

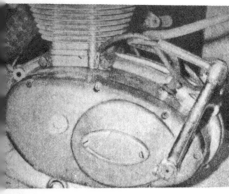
1 Note how thoroughly the engine unit has been cleaned—this is easy with such a neat design as this Ducati

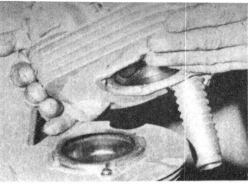
2 Undo the four head retaining bolts and lift off the head complete with the cam drive shaft and top bevel

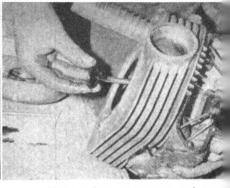
3 Using a hair spring type valve compressor, prise out the cotters and then remove the valves themselves

7 Unscrew the eight allen screws which hold the primary drive cover and then use the special extractor shown

8 Lift off the outer main shaft bearing and undo the six clutch adjusting screws in order to lift out plates

9 Turn down the lock washer and undo the clutch centre nut. You will need a special clutch holder tool to do

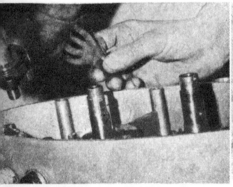
13 Remove the pressure springs—careful—and pull out kickstarter shaft and quadrant. Then remove starter gear

14 To draw out the clutch thrust rod you may have to tilt the engine over. Take care of the ball at each end

15 The gear selector mechanism is contained in this one unit and held on to the box by six allen screws

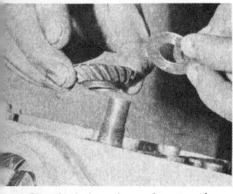
19 Free the lock washer and remove the left hand thread nut holding the bottom bevel. This has timing marks

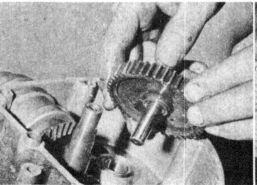
20 Then lift out the timing idler, which also has timing marks on it. These look like small holes in the metal

21 Undo the screws holding the crankcase halves together and separate them. Note the full flywheel

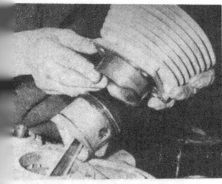

4 The barrel can now be lifted off as the head bolts pass right through holding both to the crankcase

5 Remove the gudgeon pin circlips and push out the gudgeon pin. It may need a slight tap to start it off

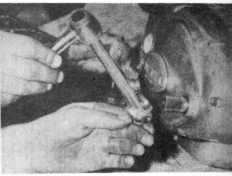

6 Next take off the kickstart lever. This is held to the gearbox cover by a pinch bolt. Clean on reassembly

10 This will allow the clutch centre and the drum to be pulled off. Check for wear on the clutch drum slots

11 Release the locking washer on the mainshaft drive pinion, undo nut and remove pinion noting keyed to shaft

12 The magnetic flywheel can now be removed with the aid of another extractor. Ignition is by E.T. system

16 Bend down the tab washer on the gearbox sprocket nut and remove the nut whilst holding the sprocket as shown

17 Next remove the contact breaker inspection plate. Two screws and bolts hold the clip on to the cover

18 The timing cover is held on by seven allen screws. This is one of the good points about this machine

22 Remove the complete flywheel assembly from the crankcase half. This may mean a slight tap with a hide mallet

23 There is a plug fitted in the flywheel so that sludge from the oilway can be cleared without stripping the big-end

24 The timing marks can be seen quite clearly in this shot of the top bevel. Note yet another lock washer

Left: Rocker-boxes are integral with the cylinder head. Note the ready access for valve-clearance adjustment. Right: Rear-suspension setting can be made without tools

249 c.c. Ducati Daytona

Businesslike High-performance Sports Model with Excellent Roadholding, Superb Brakes and Exceptional Finish

ONE'S first glimpse of the Ducati Daytona gives the impression of a beautifully balanced, stylish, sporty little mount which is guaranteed to appeal to the young blood and old hand alike. It looks exciting—and it is. A comfortable 80 m.p.h. maximum, stamina to cruise all day at 70 plus and roadholding and brakes straight out of a racing stable establish the Daytona as one of the leaders in a highly competitive class.

The engine excels in providing punch at both ends of the r.p.m. scale. From tickover speeds, it produces smooth, progressive power right up to maximum speed. Evidence of the excellent bottom-end pulling came in an effortless two-up (23½ stones) standing start on a 1 in 6 gradient. Second gear was notched in ten yards with the engine picking up rapidly all the way.

Acceleration is outstanding for a two-fifty. From low speeds a tweak of the throttle brings an instantaneous pink-free response and sends the speedometer needle flickering rapidly round the dial. Excellently chosen gear ratios encouraged snappy getaways and normal upward changes were made at 20, 35 and 55 m.p.h. If necessary the changes could be delayed for higher speeds, as the performance panel suggests.

Under average conditions, the Daytona would cruise indefinitely at 70 m.p.h. on about half throttle with the rider adopting no more than the half-crouch enforced by the clip-on bars. Given slightly favourable circumstances, speeds of well over 80 m.p.h. were possible.

Commendably quiet on small and medium throttle openings, the exhaust assumed a slightly flatter and louder note when the engine was working hard, but never became offensive.

With the carburettor lightly flooded, the air lever closed, the spring-loaded ignition switch turned on and a mere whiff of throttle, the engine usually responded to the first prod on the kick-starter. (As the crank is on the left it was more convenient to stand alongside than to straddle the seat; and the prod needed to be hefty to overcome compression.) When warm the engine would settle down to a reasonably slow, reliable tick-over. Mechanical noise was limited to faint piston slap when the engine was cold and slight valve-gear chatter when hot.

It was advisable to free the clutch plates before starting the engine; then bottom gear could be engaged with no more than a click. Heavier than average in operation, the clutch took up the drive smoothly and, even after a succession of full-throttle starts when obtaining the performance figures, there was no suggestion of plate-swelling. Of the heel-and-toe variety but operating in the usual British manner—that is, down for upward changes and vice-versa—movement of the gear pedal was short, crisp and light. Very rapid changes could be made in either direction without clashing the gears.

The sporty appearance is obtained at some sacrifice to riding comfort. Short and narrow, the clip-on handlebars are set rather low and the welded-on clutch and front-brake lever pivot blocks cannot, of course, be adjusted for position. To match the forward lean necessary with the handlebars, footrests set slightly more rearward would be an advantage for tallish riders. The seat, reasonably wide and long enough to accommodate a passenger, was harder than average but a very good point is the two-position lugs provided for the pillion footrests.

Bend swinging gave immense pleasure. Racing heritage was never more apparent than in the taut steering of the Daytona and the effortless manner in which it could be canted this way and that. It clung leechlike on line, irrespective of bumps or sharp undulations. The steering damper was superfluous. There seemed to be no limit to how far the model could be heeled over and even enthusiastic cornering failed to ground either the footrests or the centre stand. The roadholding inspired confidence no matter how greasy the surface.

Travel of the front fork was longer and softer than usual for an Italian lightweight; a shade more damping would have eliminated a suggestion of pitching experienced occasionally and probably prevented bot-

Emphasizing the sporting character of the 249 c.c. Ducati are low-set, clip-on handlebar stubs

ROAD TESTS OF NEW MODELS

toming when the front brake was applied very hard. Rear suspension was beyond criticism; a hinged clip on the units makes it possible to change the settings—there are three—without tools. The "medio" (middle) position was found to be satisfactory for both solo and two-up riding.

Extremely powerful but easily controllable, both brakes were even more potent than the figure in the performance data suggests. The front, especially, possessed real bite and the tyre could be made to squeal at will. Both brakes kept their bite in heavy rain.

The main beam of the headlamp with its 25/25-watt bulb was not intense enough for high-speed cruising on unlit roads and, although the anti-dazzle cut-off on dipped beam was excellent, the amount of light was inadequate for average British roads.

Commendably light weight (280 lb) and low centre of gravity made it possible to manoeuvre or lift the model easily but to some extent these good points were offset by a narrowish steering lock.

Maintenance necessary during the test was negligible. But all the usual jobs could be carried out easily. Accessibility is excellent. Typical of the attention to detail are the spring-loaded finger adjusters for the clutch and front-brake cables and the spring-loaded friction screw for the twist-grip.

The battery and rectifier are neatly concealed beneath the seat and the streamlined tool box on the left is matched by another box which houses the air filter on the right. An interesting feature which stops messiness is a crankcase breather venting to the air filter box. In 1,400 miles there was not a trace of oil anywhere on the machine.

Polychromatic blue with silver mudguards, tank panels, and tool-box and air-filter covers, provide an extremely smart finish. Light-alloy components are highly polished and the exhaust system, handlebars, rims and many other fittings are chromium plated. As implied earlier, the Daytona is extraordinarily smart in appearance.

For the weekend dicer, a megaphone is available, price £4 5s. The only other alterations required are to fit the carburettor bellmouth supplied, increase the main-jet size and make up a shorter rear-brake pedal. The pillion rests may be used as racing footrests. In that guise the Daytona should provide an ideal mount for the impecunious racer and to be more than capable of holding its own in average company.

SPECIFICATION

- **ENGINE:** Ducati 249 c.c. (74 x 57.8mm) overhead-camshaft single. Crankshaft supported in two ball bearings; caged roller big-end bearing. Light-alloy cylinder head and barrel; compression ratio, 7.5 to 1. Dry sump lubrication; oil capacity, 3¼ pints.
- **CARBURETTOR:** Dellorto with air filter. Air slide operated by handlebar lever.
- **IGNITION and LIGHTING:** CEV alternating-current generator charging 6-volt 13-amp-hour Safa battery through rectifier. Aprilia 6in-diameter headlamp with pre-focus light unit and 25/25-watt main bulb.
- **TRANSMISSION:** Ducati four-speed gear box in unit with engine. Gear ratios: bottom, 18.44 to 1; second, 11.06 to 1; third, 7.9 to 1; top, 6.5 to 1. Multi-plate clutch with bonded-on friction linings running in oil. Helical-gear primary drive. Rear chain ½ x 5/16 in with guard over top run. Engine r.p.m. at 30 m.p.h. in top gear, 2,800.
- **FUEL CAPACITY:** 3¼ gallons.
- **TYRES:** Pirelli: front 2.75 x 18in ribbed; rear 3.00 x 18in studded.
- **BRAKES:** Approximately 7in-diameter front, 6¼in-diameter rear with finger adjusters.
- **SUSPENSION:** Marzocchi telescopic front fork with hydraulic damping. Pivoted rear fork controlled by Marzocchi spring-and-hydraulic units with three-position adjustment for load.
- **WHEELBASE:** 52¼in unladen. Ground clearance 5¼in unladen.
- **SEAT:** Giuliari twin-seat; unladen height, 29¼in.
- **WEIGHT:** 280 lb fully equipped, with full oil container and approximately one gallon of petrol.
- **PRICE:** £210 16s 3d; with purchase tax (in Great Britain only), £258 12s 11d.
- **ROAD TAX:** £2 5s a year.
- **CONCESSIONAIRES:** Ducati Concessionaires, Ltd., 80, Burleigh Road, Stretford, Lancashire.

PERFORMANCE DATA

MEAN MAXIMUM SPEED: Bottom: *32
Second: *53
Third: *74
Top: 81
*Valve float occurring.

HIGHEST ONE-WAY SPEED: 84 m.p.h. (conditions: gusty side wind; 13½-stone rider wearing two-piece suit).

MEAN ACCELERATION:

	10-30 m.p.h.	20-40 m.p.h.	30-50 m.p.h.
Bottom	4.8 sec		
Second	4.6 sec	4.6 sec	6.4 sec
Third	—	6.0 sec	6.6 sec
Top	—	6.6 sec	6.8 sec

Mean speed at end of quarter-mile from rest: 67 m.p.h.
Mean time to cover standing quarter-mile: 18.6 sec.

PETROL CONSUMPTION: At 30 m.p.h., 112 m.p.g.; at 40 m.p.h., 96 m.p.g.; at 50 m.p.h., 68 m.p.g.; at 60 m.p.h., 65 m.p.g.

BRAKING: From 30 m.p.h. to rest, 31ft (surface, dry tarmac).

TURNING CIRCLE: 20ft.

MINIMUM NON-SNATCH SPEED: 16 m.p.h. in top gear.

WEIGHT PER C.C.: 1.12 lb.

Cycle World Road Test
DUCATI SCRAMBLER

BELLA ITALIA, beautiful Italy, is the spiritual home of all the world's sporting conveyances. They discovered the joys of chariot racing thousands of years ago, and to this day Italians prefer speed on wheels to almost any other form of sport. Pass any Italian on the road and you will have a race on your hands; that is the way they think, and live, and that is the reason behind Italian machinery being like *it* is.

Right at the top of the long list of Italian two-wheelers that have captured our fancy is the Ducati. These fine small-displacement machines are made in Italy's industrial north, where technical excellence is as highly regarded as any other form of artistry. Much of this technical excellence is to be seen in the Ducati motorcycles, starting with their remarkable racing twin, in which three camshafts open and close the valves mechanically, without the aid of springs, right down to their ultra-light and economical Falcon-50, which goes so far on a gallon of gasoline that more fuel is lost through evaporation than is burned by the engine. They also have a series of 250cc "singles," which are distinguished by having an overhead camshaft and a willingness to rev, and it is the scrambler version of this bike that is the subject of this report.

We may as well start with the Ducati engine, as it is the outstanding feature of the bike. It is an all-alloy single that delivers its maximum power at 7500 rpm with the cam used in the special scrambler version. Externally, the engine (which is in unit with the transmission) presents a nice, neat appearance. Part of this

neatness is due to the total absence of oil leaks, but much of it stems from the use of recessed, allen-head screws in place of the customary cap-screws holding the assembly together. The castings are of unpolished aluminum, but they are so smooth and clean, right out of the mold, that buffing would just be superfluous. The finning around the cylinder barrel and the head is deep and, all in all, it puts up a most impressive appearance.

Inside, the engine gets even better. A fairly conventional double-flywheel, pressed together crank assembly is used, but there are such touches as a sludge extractor built into the crankshaft. The oil is fed into the timing-gear flywheel and is led out into a chamber in the rim of the flywheel before flowing back to the passage leading to the crankpin bearing. This sludge extractor is assisted by a more ordinary filter located in the bottom of the crankcase — this being a wet-sump engine.

The engine's single overhead camshaft is driven by spiral-bevel gears and a tower-shaft; the overall layout is much like that of the single-cam Norton racing engine. Short rockers transmit motion from the cam lobes to the valves, and the valves are closed by means of hairpin-type springs. The mechanism is light, and does its job effectively. The springs are entirely enclosed in the cylinder head casting and, from the copious flow of oil that is directed toward them, we would guess that oil is being used to cool the valves and springs, as well as a lubricant. It was such small touches as this, and other detail items too numerous to catalogue, that created, for us, an overwhelmingly good impression of the engine.

The primary drive from engine to clutch is a pair of helical-cut gears, which give a 2.65:1 reduction. The larger, driven gear carries the clutch. All of the transmission gears are in constant mesh and they are lubricated by engine oil. A rocker-type change lever is fitted, which means that one need not play push-pull with one's toe, and the shifting dogs all slide in and out of engagement without fuss or forcing. Indeed, we discovered that forcing the shift would sometimes ram it right past engagement and into one of the between-gears neutral positions. Practice and more restrained tactics were enough to overcome this problem.

The clutch demonstrated a demon grip throughout the duration of the test. While we were impressed by the determination it showed, we would have been just as happy if it had made its engagements without such a show of strength. It wasn't quite what we would call "grabby," but it was definitely sudden in its action. This sudden bite was to cause us some unexpectedly-exciting moments during our acceleration runs. The knobby tires maintained only a rather precarious grip on asphalt in any case, and on fast starts, the back wheel would spin like fury. Our test rider was surprised no end to travel off down the strip with the wheel slipping and the rear of the bike switching back and forth just like a big-inch dragster. Very seldom do we get that much action out of a 250cc motorcycle.

The Ducati's frame is a nice piece of work. It consists primarily of a single curved tube that arcs from the steering head, over the engine, and down to the rear suspension pivot. The structure is braced by means of a tube that connects the steering head and the front of the crankcase, which ties in the ends of the "backbone" tube and makes it the equivalent of a single-loop frame — albeit slightly lighter. Miscellaneous curved lengths of small diameter tubing hold the suspension struts, rear fender and saddle.

Nominally, this Ducati Scrambler model is a "four-in-one" machine. It comes with knobby tires and equipment obviously slanted toward off-the-road conditions, but it is also intended for road racing. A change of tires are all that is needed. So say the makers, and they are right. The Berliner Motor Corporation, who distribute the Ducati, list this model as one for street riding, road racing, short-track racing and scrambles. In all four categories it would be superb. It is fast and smooth enough for touring, and has the handling and brakes for road racing, with the high-winding power and stamina required on flat tracks and scrambles courses. However, on some cross-country courses we have seen, the Ducati would be at a disadvantage. When the going gets really nasty, the bike's wheelbase, fork angle and suspension are not precisely what one might call ideal. Although it is a whirlwind on pavement and on moderately rough dirt, those pumpkin-sized clods that are a standard feature of some cross-country courses might be too much for the Ducati's essentially road-going frame

and forks. This does not, of course, reflect unfavorably on the Ducati's overall performance. On the contrary, it is astonishing that it will do such a wide variety of things so well. Its deficiencies in rough terrain simply reflect the fact that no bike will do everything equally well.

Starting these small, high-compression singles can be a real chore, and we were pleasantly surprised to find that the Ducati would come to life with very little urging. The ignition has an automatic spark advance, which keeps the engine from biting back, and the engine proved to be a one-kick starter once it was warm. From cold, a trifle more effort was required, but not enough to tire us.

The brakes were obviously intended for road-racing, as they are very large, with finned aluminum drums, and air-scoops and vents cast into the front backing-plate. With so much brake area, and the speed of our test bike limited so severely by its gearing, we never managed to work them hard enough to get them warm — much less produce any fade. Furthermore, we haven't heard any complaints about their effectiveness from the men who are road-racing Ducatis, and winning.

Appeal of the Ducati Scrambler is further enhanced by the lavish selection of gears and extras that are available. They include: a full set of control cables, valve adjustment caps, a pair of rigid frame members to replace the rear shocks for use on a flat track, tachometer drive unit, three extra rear sprockets (45, 50 and 60 tooth), and an extra 16 tooth countershaft sprocket. By the simple expedient of experimenting with this huge selection of ratios one can either push the top speed up to 100 mph, employ a strain-free overdrive, or set up for maximum (and a bit shattering) acceleration.

Our test machine was equipped with the standard 55 tooth rear sprocket and 14 tooth countershaft sprocket, both of which are somewhat of a compromise yet offer amazing performance for a 15 cubic inch cycle.

The finish on this Ducati scrambler is extremely good. Parts are made with obvious care, and fitted with precision. The paint was much more restrained in color than any previous Ducatis we had seen, being a light pearlescent blue, and we thought it looked great. Whatever its short-comings out in Farmer John's plowed "south-forty," it more than compensates for this with its speed and handling every place else. If you are in the market for a 250, even a cross-country bike (you can always juggle the fork angle), don't miss this one. •

DUCATI 250 SCRAMBLER

SPECIFICATIONS

List Price	$669
Frame Type	tubular, single-loop
Suspension, front	telescopic fork
Suspension, rear	swing arm
Tire size, front	3.00-19
Tire size, rear	3.50-19
Brake lining area, sq. in.	253
Engine type	single-cyl, sohc
Bore & stroke	2.92 x 2.28
Displacement, cu. in.	15.2
Displacement, cu. cent.	248.6
Compression ration	9:1
Bhp @ rpm	30 @ 7500
Carburetion	27mm (1.06") Dellorto
Ignition	flywheel gen. & coil
Fuel capacity, gal.	3.4
Oil capacity, pts.	5.1
Oil System	wet sump
Starting system	kick, folding crank

POWER TRANSMISSION

Clutch Type	multi-disc, oil-bath
Primary drive	helical gear
Final drive	single-row chain
Gear ratio, overall:1	
4th	7.51
3rd	8.91
2nd	12.8
1st	21.3

DIMENSIONS, IN.

Wheelbase	52.0
Saddle height	28.3
Saddle width	9.5
Foot-peg height	10.5
Ground clearance	6.5
Curb weight, lbs.	277

PERFORMANCE

Top speed, average	82
best run	82.0
Max. safe speed in gears @ 8400 rpm	
3rd	72
2nd	50
1st	30
Fuel consumption range, mpg	n.a.
Mph per 100 rpm, top gear	10.2

SPEEDOMETER ERROR

30 mph, actual	no speedometer
50	
70	

ACCELERATION

0-30 mph, sec.	4.1
0-40	6.0
0-50	8.7
0-60	11.5
0-70	16.0
0-80	22.5
0-90	
0-100	
Standing ¼ mile	17.3
speed reached	73

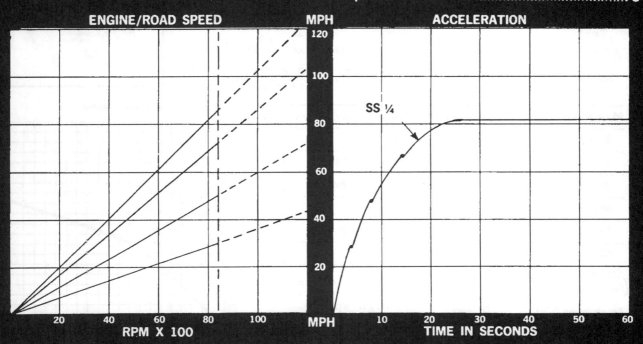

DAVID WEIGHTMAN TESTS THE DUCATI DAYTONA

ROAD TEST

Exciting performance from this 250 c.c. Italian sport lightweight motorcycle..

'M.M.' MACHINE TEST REPORT No. 44

Machine: Ducati   Model: Daytona   c.c. 248
Test Mileage: 800   Price new: £254.6s.1d.   Examiner: D. Weightman
Supplied by: Ducati Concessionaires Ltd., 30 Burleigh Road, Stretford, Lancs.

Maximum Points 10 (compared with machines in same c.c. and price range)

Brakes (front)	10	Lights	8
Brakes (rear)	9	Engine accessibility	8
Brakes (both)	9	General performance	9
Steering at high speed	10	Overall finish	9
Steering at low speed	9	Electrical layout	8
Gearbox action	10	TOTAL	99

Overall Fuel consumption: 70 † m.p.g.   Remarks: † Mainly at high speeds
Acceleration 0-50: 8 secs.
Top speed: 86 m.p.h.

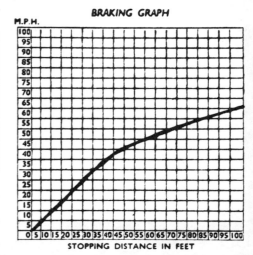

BRAKING GRAPH
STOPPING DISTANCE IN FEET

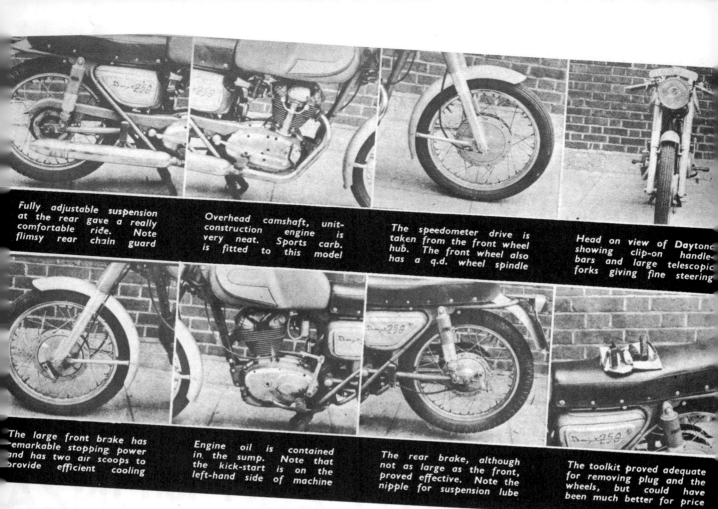

Fully adjustable suspension at the rear gave a really comfortable ride. Note flimsy rear chain guard

Overhead camshaft, unit-construction engine is very neat. Sports carb. is fitted to this model

The speedometer drive is taken from the front wheel hub. The front wheel also has a q.d. wheel spindle

Head on view of Daytona showing clip-on handlebars and large telescopic forks giving fine steering

The large front brake has remarkable stopping power and has two air scoops to provide efficient cooling

Engine oil is contained in the sump. Note that the kick-start is on the left-hand side of machine

The rear brake, although not as large as the front, proved effective. Note the nipple for suspension lube

The toolkit proved adequate for removing plug and the wheels, but could have been much better for price

DUCATI motorcycles are renowned for their powerful little engines, but I didn't realise that these machines were just as good in other respects. It therefore came as a very pleasant surprise to find that the rest of the 250 c.c. Daytona was up to the same high standards as the overhead camshaft motor.

The handling of the machine was really superb. The bike could be flicked over for a bend with practically no effort at all and would stick exactly to the line chosen. There was never any suggestion of "wag" on leaving a corner and the thing seemed almost to pick itself up as you accelerated out. "S" bends could be thoroughly enjoyed with no fears of overdoing it, and this applied throughout the speed range and not just at low speeds. Even at 80 m.p.h. the bike corners in such a way that every rider feels like an expert.

The next surprise was the brakes. We have had several machines in the past few months which have had really first-class brakes, and the Daytona is certainly worthy of addition to the list. Although they do not have twin leading shoes, these brakes really stop you from all speeds without fade or judder. Only a moderate amount of pressure was needed on the levers, and it seems as if Ducati have done the impossible in providing a brake which works well at both low and high speeds.

It's just as well that the handling and brakes are as good as this because the speed of the bike certainly makes it necessary. The overhead camshaft engine pushes the bike along at 85 m.p.h. in standard trim. If you put the racing kit on, the bike will do around the "ton" with a silencer, and over 105 m.p.h. with megaphone. This is motoring for a 250!

The racing kit consists of a 9:1 piston, 27 mm. sports carb., assorted jets, two rear wheel sprockets and a megaphone. This is available from the concessionaires at a price of £18 10s.

However, back to the engine. With this sort of performance one does not expect a quiet machine and yet the Daytona exhaust note was surprisingly quiet. The bike only seemed noisy to the rider, who heard all the mechanical clatter of the overhead camshaft valve gear. There was remarkably little vibration.

Good Gearbox

The gearbox deserved top marks too. The change was one of the best I have ever used, and the rocking pedal is the only one on which I have been able to actually use the heel part comfortably. Both upward and downward changes could be made without using the clutch and the gears always engaged quickly and quietly.

Clutch action was very light, despite the fact that the clutch was made for competition use. To say that it was either in or out would be an exaggeration, but the take up was pretty rapid. But it does take hard use without complaining and is quite easy to operate.

Lighting on this model was only average but the battery was not at its best and I think that with a new one the lights would have been very good. The horn, too, was not as good as it should have been. The horn and dipswitch unit was not well designed and was also very difficult to operate.

General comfort was pretty good for fast riding, but the position was much too crouched for town work. This was aggravated by the fact that the handlebars and levers were not fully adjustable, and this meant that my hands were in a twisted position for town use. The dualseat was surprisingly comfortable and even with the firm suspension gave quite a reasonable ride.

A surprising thing about this sports machine was the ease with which it could be started. Almost always the first kick would be enough and the tickover was quite reliable too. Fuel consumption was reasonable at 65-70 m.p.g., and even if the bike was thrashed it didn't drop much below this figure.

For motorcyclists with sporting instincts, this is a very fine machine which isn't too thirsty. If you want to go racing, you could hardly do better at the price, because the performance with the racing kit on is quite something. I know—I tried one!

HOW LONG DOES IT TAKE?

Adjust rear chain	1½ minutes
Adjust both brakes	½ minute
Adjust tappets	12 minutes
Adjust points	1½ minutes
Top up battery	4½ minutes
Remove and replace rear wheel	4½ minutes

DUCATI 250 MONZA

A VERY BIG NAME in Italian motorcycling is Ducati, and it is a big name for an excellent reason: Ducati's machines are among the best engineered and best finished to be found anywhere in that country. Gilera and MV get most of the glory, with their racing "fours," but it is Ducati that produces a top-flight motorcycle for the man on the street.

The Ducati that is the subject of this report is their 250 Monza — named for the famous Italian race course. It has a single-cylinder engine, with a short stroke to give low piston speeds and a big bore to allow for maximum valve size. The valves are operated by a single overhead camshaft, with rockers (that double as cam followers) carrying the motion from the cams to the valves. There is little "monkey-motion" in the valve gear, and not much reciprocating weight, with the result that valve-float does not occur until the engine is forced far past its power peak — an act of foolishness that few riders would commit.

The camshaft is driven through spiral-bevel (for quietness) gears and a tower shaft — and the tower-shaft housing also serves as an oil return from the cylinder head. This cylinder head is, like most of the engine, of aluminum, and deeply finned except around the valve-chest, which appears to depend on oil for its cooling. The hairpin type valve springs are completely enclosed, with finned, removable covers that give access to the springs and the screw-type valve clearance adjusters. On both of the Ducatis we have had for testing, the valve gear was quiet and not even oil seepage — much less leaking — ever appeared around the cylinder heads.

Bottom-end components in the Ducati engine are much like any other single: roller and ball bearings throughout. One touch we have mentioned before, and like, is the inclusion of a sludge-trap in one of the flywheels. Oil being pumped to the bearings has to flow part-way out into a pocket in the flywheel (located opposite the connecting-rod) before traveling back across to the rod bearing. Any particles in the oil are caught by centrifugal force and packed out into the sludge trap, which keeps the oil a lot cleaner than it would otherwise be. Of course, the sludge trap is fitted with a threaded plug, which may be removed for purposes of cleaning when the engine is apart for repairs. And, a wire-mesh filter (removable) is included at the oil pick-up in the sump to strain out any of the coarser particles.

Drive to the clutch is through a pair of 2.5:1 helical gears, the larger of which carries the clutch. No shock absorbing device is provided — and apparently none is needed, for the unit feels very smooth. From the clutch the drive goes into an all-indirect transmission that is mounted in the back of the crankcase and lubricated by engine oil. All gears are in constant mesh and gear selection is done by sliding dogs.

The engine/transmission unit has all aluminum casings and even the cylinder barrel is of aluminum, with an iron liner. The various castings are unpolished, but as they are all die-castings, and very smooth as pried from the molds, polishing is actually quite unnecessary. Recessed allen-head cap screws are used to hold things together, and the entire unit is extremely clean and attractive.

We have occasionally been asked what happens when someone tries to slip us a "ringer;" that is to say, a bike that has been fiddled to make it go faster. Now, everyone knows: we take note of the modifications and report on them. This happened on the Ducati Monza supplied to us for this test; the dealer that prepared the bike, in a poorly considered fit of enthusiasm for the product, carved out the ports and replaced the standard 24mm carburetor with one having a 30mm throat. He also replaced the stock gearing, which gives 6.42:1 as an overall reduction, with sprockets that gave 7.26:1. The gearing change gave the bike good acceleration, but limited the top speed (the machine was running flat-out and at its peak just before the ½-mile distance was reached on our "practical maximum" run). As for the carburetor; we are not convinced that it did any good at all. The cams had not been changed, and the engine definitely felt a bit over-carbureted: it popped and sputtered at low speeds and felt no more than stock-strong at any speed. Thus, we are of the opinion that the figures given on our data page represent fairly closely what one might expect of a Ducati Monza with a stock engine, but with slightly altered gearing.

The bike shown in the photographs on these pages is not our test machine, but was loaned to us for photographic purposes by Nicholson Motors in North Hollywood, California, due to the non-standard condition of our actual test machine.

One of the interesting things about the Ducati was that even with that huge carburetor, it was very easy to start— hot or cold. The kick-lever is on the left-side of the bike, and that was a confounded nuisance for our predominantly right-footed kickers, but not many whacks at the lever were needed to make the engine start. Of course, there was a marked reluctance to run smoothly until the engine was thoroughly warm (that huge carburetor again) and it was necessary to keep the air-bleed slide closed until it came up to temperature.

The Ducati's frame and suspension system are nothing very much out of the ordinary — but they do give extraordinary results. The frame is made of steel tubing, with a main member that forms a curved backbone from the steering head to the rear of the engine/transmission casing, where the rear suspension's swing arm pivots. There is also a down-tube to the front of the engine, and the lay-

out gives the equivalent of a single-loop frame. Enough clearance has been provided around the top of the engine to permit easy servicing in that area.

No special feature distinguishes the front forks, but as we said, they give extraordinary results. The springing is soft, and the damping very firm, and the combination of rake and trail must be good, for no tendency toward fork-flutter ever appeared. A steering damper is provided, but the only function it apparently serves is to keep the steering from being unduly sensitive. The steering is very precise, probably because the forks have been made exceptionally rigid. The axle is clamped very solidly, which lends rigidity to the forks as a whole, and there is a heavy strap bridging between the moving legs that serves as a fender bracket and as further reinforcement for the forks themselves. As elsewhere on the Ducati, there is a considerable amount of light alloy in evidence.

Full-width, light alloy brakes are a feature, and they are light and powerful in action — being, at times, almost too powerful if the rider clamps them on with more vigor than finesse. Also, they had a rather spongy feel that made it somewhat difficult to sense how much braking force was being applied. However, we used them quite hard for a number of laps around the difficult and demanding Riverside Raceway and they gave no trouble except when used clumsily — and it can be argued, with some logic, that such things are a problem for the rider; not for the people who make the machine.

A good, comfortable riding position, in combination with a good, soft seat, made riding the Ducati a special pleasure. The handlebars are not much for flat-on-the-tank racing, but are excellently arranged for touring, and the various control levers, buttons and switches are located near to hand. The only thing we did not care for was the fact that the clutch and brake levers did not have ball-ends (a recent wounding has made us sensitive on this point).

The gear-shift control is a rocker-type pedal, which we prefer to the single-lever, toe-lift variety, and it gave clean, positive engagements — as it should. Some trouble was experienced, at first, with the transmission's unwillingness to dab into top gear as neatly as it did into the other three, but we soon learned to apply more pressure and that ended the difficulty. The transmission ratios are especially well staged; a trifle wide for racing, but perfect for touring.

Speedometer error was very great, but we have reason to believe that this is not typical. Actually, the instrument was in the process of breaking even as we made our speedometer calibration runs, and before the testing was completed, it had completely fallen apart. While it worked, it gave steady, if rather misleading, readings.

One item that we disliked so much that we removed it was the lean-stand. The Ducati is fitted with a very convenient and easy to operate "rock-back" stand, and that was fine, but it also had a swinging prop that would have also been fine except for one small detail: it extends down and outward far enough to "ground" every time the machine was banked over very far in a left turn. This was particularly outrageous in view of the fact that the Ducati will corner so well. After removing the stand, we could drop over in a bend until the pegs were about to drag (you "feel" the pavement with the side of your boot-sole to make certain things do not go too far).

For all around sports/touring/racing use, it would be hard to find a better bike than the Ducati. Its finish is marvelous, it is quite fast, and it handles in a way that few others can match. Also, it is so smooth that one begins to wonder if there is any point in building twins (we know there is, but not for reasons of smoothness) and we have had good reports of its durability. The makers offer the machine in several forms — depending on the use for which it is intended — and they have just come up with a super-tuning kit (consisting of a cam and special piston) for those who like that sort of thing. Try the Ducati and you will like it; we did. •

DUCATI MONZA
SPECIFICATIONS

List Price	$579.00 FOB, L.A.
Frame Type	tubular, two-loop
Suspension, front	telescopic fork
Suspension, rear	swing arm
Tire size, front	2.75-18
Tire size, rear	3.00-18
Brake lining area, sq. in.	n.a.
Engine type	single cyl, sohc
Bore & stroke	2.92 x 2.28
Displacement, cu. in.	15.2
Displacement, cu. cent.	249
Compression ratio	8.0:1
Bhp @ rpm	(stock) 24 @ 7500
Carburetion	30mm (1.18") TT Dellorto
Ignition	battery and coil
Fuel capacity, gal.	3.4
Oil capacity, pts.	5.1
Oil system	wet sump
Starting system	kick, folding crank

POWER TRANSMISSION

Clutch Type	multi-disc, oil bath
Primary drive	helical gear
Final drive	single-row chain
Gear ratio, overall:1 (non-standard)	
4th	7.26
3rd	8.85
2nd	12.4
1st	20.6

DIMENSIONS, IN.

Wheelbase	52.0
Saddle height	31.5
Saddle width	4.0
Foot-peg height	9.25
Ground clearance	5.0 (at stand)
Curb weight, lbs.	292

PERFORMANCE

Practical maximum speed (after ½-mile run)	77
Max. speed in gears @ 7800 rpm	
4th	77
3rd	64
2nd	45
1st	27
Mph per 1000 rpm, top gear	9.9

SPEEDOMETER ERROR

30 mph, actual	22.9
50	42.0
70	58.8

ACCELERATION

0-30 mph, sec.	3.8
0-40	5.8
0-50	8.6
0-60	11.4
0-70	16.8
0-80	
0-90	
0-100	
Standing ¼ mile	17.4
speed reached	71.5

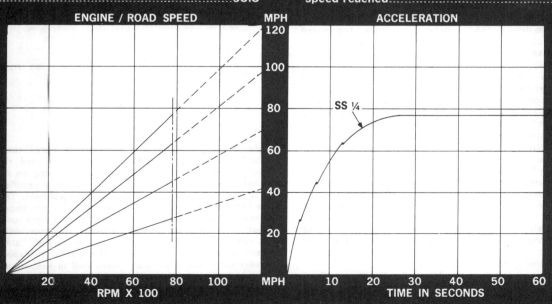

Road impressions of new models

DUCATI 48 C.C. SPORTS

by MIKE EVANS

DUCATI lightweights have proved themselves over many years on the British market—and now they have a greenhorn in their midst: a two-stroke fifty! Is it a worthy partner of the others? The answer is an undoubted yes.

It *looks* like many Italian fifties and yet it has charm and character all its own. And at the price it is excellent value for money.

On the engine department the machine could not be faulted. The sporty power egg was a willing worker which gave an exhilarating performance when buzzed.

The gear change—through a twistgrip control—was slick in action and the bike would run up and down the scale with great verve. A disadvantage to such enthusiastic use, though, was the decidedly fruity exhaust note; this prevented full performance being used in town.

Speeds in the gears were 20 and 30 m.p.h. in bottom and second respectively. The jump from second to top was rather excessive and a change was always accompanied by a noticeable drop in revs.

Allowing for slight speedometer inaccuracy, top speed was just over 50 m.p.h. Cruising speed was about 40 m.p.h., given neutral conditions. And this for a fuel consumption of well over 100 m.p.g.

Slight vibration was transmitted through the handlebars and footrests, becoming more pronounced as the revs rose. However, it was not severe enough to be annoying.

STABILITY

Starting was easy once the correct drill was known. When cold, best results were obtained by flooding the carburettor and depressing the air lever before kicking.

The engine was easy to flood because of the extreme downdraught of the carburettor. This had to be kept in mind, especially if the machine was to be parked for a few minutes, when it was imperative to turn off the fuel tap.

Handling was faultless. Similar fifties all seem to have inherent stability. Cornering the low, light sportster could leave no one in doubt that riding a Ducati .48 is great fun.

When delivered the clip-on handlebars touched the tank with the steering on full lock. This was easily rectified, however, by adjustment. The riding position, too, was improved.

All controls were conveniently placed (the rear-brake pedal is on the "British" side). The brakes were a trifle spongy, but were well in keeping with the machine's performance. Lighting, too, was up to standard.

Specification

ENGINE: Ducati 48 c.c. (38 x 42 mm) two-stroke single. Compression ratio, 9.5 to 1. Petroil lubrication; ratio, 16 to 1.
IGNITION and LIGHTING: CEV flywheel magneto. Direct-lighting coils. Twin-filament, six-volt 15/15w main bulb.
TRANSMISSION: Three-speed gear box with twist-grip control. Primary drive by helical gears, secondary drive by chain.
FRAME and SUSPENSION: Duplex cradle frame. Hydraulically damped telescopic front fork and pivoted rear fork.
TYRES: Pirelli 2.25 x 19in front and rear.
WEIGHT: 119 lb without fuel.
ROAD TAX: £1 per year.
PRICE: £89 19s 6d (including British purchase tax). Price includes speedometer.
CONCESSIONAIRES: Ducati Concessionaires, Ltd., 80, Burleigh Road, Stretford, Lancs.

TOUR TEST —
TRAILERING WITH THE DUCATI BRONCO

by Cliff Boswell

We loaded the little Ducati 125cc motorcycle aboard our panel truck, hooked on the trailer and left for points east and north. Two months and 6,400 miles later we unloaded at the starting point. Surprisingly enough, the Ducati's odometer showed 850 miles although the bike rode piggy-back the entire distance. Therein lies a tale.

Ever since attending an international trailer rally at Auburn, Washington, in 1962 I had toyed with the idea of carrying a light motorcycle on trailer trips for auxiliary transportation. There were perhaps a hundred or more motorcycles and scooters present at the 1962 gathering, and needless to say, their owners were enthusiastic about the handy transportation these machines afforded for running into town or for sightseeing. Therefore, when I decided to attend a Wally Byam Caravan Club rally in Bemidji, Minnesota, and to join a caravan to Canada afterwards, I knew I had to have a motorcycle.

Obviously, my BMW R60 was too heavy; so a Ducati Bronco was chosen as best suited to my particular needs. A combination of tire size, weight, motor capacity, general appearance and price sold me.

The first real work-out for our little Bronco came in the Black Hills of South Dakota when 16-year-old son, Carl, rode it from our campsite in the Custer State Park to Mount Rushmore and other places of historical and scenic interest nearby. We had loosened it up with 300 miles of highway and broken field riding before beginning the trip; so the bike took to mountain grades in the Black Hills with agility.

Before our readers make wrong conclusions, though, I must point out that this cycle is not suitable or designed for fast highway travel, as its cruising speed on the level is about 40 mph with a top speed of 55 or thereabouts. With standard gearing it is great for twisting mountain grades where speed is not desirable, and for dirt and gravel roads, open meadows, trails and city streets. With a suitable front fender (shorter), skid plate and larger rear sprocket, the Bronco could be a wonderful trail machine as well.

An objectionable feature, in my estimation, is the high gear ratio in first gear which necessitates slipping the clutch excessively when starting. The ratio is so high that the motor is strained to put the machine into motion, and a resulting clanking noise somewhere in the gears or motor is very obvious until some momentum is reached. To an owner of a new machine this is a rather disagreeable and disconcerting distraction.

This is not the only machine that I have observed with similar gearing, and each time I ride one my reaction is the same — my estimate of the motorcycle automatically drops about 20 percent. There must be some reason for the high ratio, but I cannot think of it. With a four-speed gear box (the Bronco has one) low gear should be designed to get the vehicle going without undue clutch slippage and motor strain.

Now to go on to more pleasant thoughts.

Our next stop was in Deadwood, South Dakota, where Wild Bill Hickok met his doom in August 1876. Hickok and Calamity Jane, another legendary figure of the west, are buried in a cemetery overlooking the town.

Deadwood, and an adjoining city, Lead, are noted for having developed into gold-rush towns considerably after the stampede to California in 1849. The biggest gold mine in the world is located at Lead (pronounced Leed). Both places are rich in historical interest, and Deadwood, especially, boasts a background of bristling gunfights among claim-jumpers and between Indians and settlers that is second to none in the west.

We used the Bronco for sightseeing trips around here, sometimes riding it double. Part of the main section of Deadwood was barricaded because of road construction, but no one objected to our riding the little motorcycle along the sidewalks — in fact, people got quite a kick out of it.

Carl wheels the little Bronco into our panel truck using a board of ¾" plywood as a ramp. Scene is in the Black Hills of South Dakota.

To visitors to Deadwood I can recommend the drama "The Trial of Jack McCall" for an evening of top entertainment. McCall was the outlaw who murdered Wild Bill, and the play recounts his trial. It is enacted by local talent.

The Bronco fully came into its own, however, after reaching Bemidji, where more than 2,000 Airstream trailers were gathered at the fairgrounds for an international rally. The huge assemblage spilled over to include three adjacent fields, and the distance from the main entrance to the farthest trailer was about three-quarters of a mile. We were parked one-quarter mile from the center of activities. A distance of two miles intervened between the fairgrounds and the principal section of Bemidji.

One of the first duties taken on by the Ducati was that of delivering messages during a designated time each day. Many other small motorcycles were involved in similar activ-

A marker at the headwaters of the great Mississippi River in Minnesota.

Fearsome countenances of Paul Bunyan and his blue ox, Babe, dwarf the Ducati Bronco at the lakefront at Bemidji, Minnesota, home of the famous Paul Bunyan legend.

ities, but none performed more efficiently or evoked greater favorable comment from caravanners.

Other common chores for the Bronco were to run in to town for groceries or to amble down to the lake for photographs or to take Carl swimming. Almost every teenager at the rally, and there were many, rode the Ducati at one time or another.

I rode it in the 4th of July parade at the rally site and along the 2½-mile parade route downtown. With an American flag flying from its handlebars it joined about 30 other motorcycles, including a team of performing Hondas and formation-riding scooters, to give the holiday crowd a real kick. Northern Minnesota sees few motorcycles because of its short riding season, but residents were nonetheless enthusiastic about the showing of our little bikes.

An interesting side trip with four Honda trail bikes, the Ducati, a foldup Centaur scooter (we called it an animated suitcase) and a front-wheel driven Solex developed when some of us decided to ride the back trails and roads. This country is quite flat with some rolling knolls and many shallow lakes interspersed with clumps of small conifers and light brush. Many little-used roads and trails add to the lure of these backwoods.

The fascinating thing about our ride was not that the Ducati performed well, but that the Centaur, with barely four inches of clearance, and the Solex, with its miniature motor located above the front wheel, were able to keep up. I still can hardly believe it.

After the Bemidji rally we loaded the Bronco into the truck and joined 105 other Airstreams in a caravan across Canada. The border was crossed at Emerson Junction, and we proceeded to Winnipeg for our first stop. Unlike Mexico, no vehicle permit or tourist card is required in any part of Canada. A minimum of red tape is encountered at the border, no tipping is expected and insurance bought in the U.S. is legal. The only official document issued is a wallet-size green card for each vehicle. It is released to border officials at the port of departure.

No declaration of cameras and other equipment was required. The Canada-U.S. border is completely unguarded and unfenced. It is sad that the same cannot be said for the Mexico-U.S. border.

From Winnipeg, Manitoba, the caravan proceeded by easy stages across Saskatchewan, Alberta and British Columbia, stopping several days in Regina, Calgary and Banff National Park. At every stop the

A tour to points of interest around Banff, Alberta was made on the Bronco. This is the famous hotel there.

A Honda 125 and a tandem wheeled trail machine flank the Ducati in the 4th of July parade at the Bemidji fairgrounds. Berets were worn by all caravanners.

Bronco was wheeled out, and many were the compliments of Canadians regarding its quiet motor and its trim appearance.

We left the caravan at Golden, British Columbia to head southeast across Montana to Yellowstone and Grand Teton National Parks. From there we progressed westward across Idaho to Oregon, then swung south to eastern California and, eventually, home on the California coast. I was surprised to see an identical machine at a motel in a remote section of Wyoming. This particular model has not been on the market long, but its popularity should grow rapidly.

Among the many desirable features of the Bronco are a 3½-gallon gasoline tank (good for about 300 miles), sturdy front and rear brakes, magneto ignition, battery-powered lights and gears instead of a front chain to transmit power from motor to transmission. The latter is made possible through a unit construction of gearbox and crankcase. Overall appearance is one of precise Italian craftmanship. It is a well-proportioned, beautiful little machine.

46 M.P.H. – 105 M.P.G. – FROM THE SPORTY LITTLE ITALIAN
DUCATI 50

MOTORCYCLE MECHANICS ROAD TEST No. 67   Price new £89. 19s. 6d

- **Vehicle:** Ducati Sport 48
- **Engine:** 48cc two-stroke 4.2 bhp at 8600 r.p.m
- **Gearbox:** three-speed, twist-grip handchange
- **Final drive:** Chain

GENERAL INFORMATION
- Weight: 119 lbs
- Saddle height: 2ft. 6ins
- Turning circle: 11ft 2ins
- Is toolbox lockable: No
- Is steering lockable: No
- Fuel tank capacity: 2.2 galls
- Reserve capacity: approx. 1pt
- Oil tank capacity: —
- Gearbox capacity: approx ½ pt
- Fuel specified: two-stroke mix
- Overall consumption: 105
- Braking from 30 mph: 31 ft
- Acceleration 0–40 mph: 14 secs

EQUIPMENT SUPPLIED
STANDARD FITTINGS: Clip-on handlebars, racing style dual seat

OPTIONAL EXTRAS: Fairing, touring handlebars

SPARES PRICES
- Engine gasket set: 5s 0d
- Set valves & guides: —
- Piston with rings: £1 12s 0d
- Set of clutch plates: 13s 1d
- Silencer: £2. 13s. 0d
- Pr Exchange brake shoes: 10s 0d

RATING CHART (points out of 10)
Item	Points
Control positioning and adjustment	6
Extra Instruments and equipment	0
Fuel reserve and tap operation	8
Ease of starting	8
Engine smoothness	9
Quietness of engine & transmission	6
Gearbox and clutch operation	7
Road holding	8
Braking efficiency	8
Comfort and ease of handling	7
Lighting efficiency	6
Stand operation	6
Tool kit	4
Overall finish	8
General performance & reliability	8
TOTAL (maximum points 150)	**99**

DUCATI is a name associated with sporty Italian lightweight motorcycles. Normally, they are overhead camshaft four-stroke models, but the baby in their range, the Sport 48 is the exception to the rule. It is powered by a 48 c.c. two-stroke 3-speed gearbox unit construction engine.

However, Ducati appear to make two-strokes as excellent as four-strokes, for the power produced from the egg-cup sized unit is quite surprising. The motor itself is quite flexible and the machine may be driven at walking pace in first gear.

Similar to many fifties, one's first impression is of instability. The steering seems exceptionally light and one tends to wander off course when caught by an unexpected side-wind. Bumps in the road also give a similar sensation. However, once one becomes accustomed to the lightness of the machine, it is possible to run rings around the larger capacity models in traffic and on corners.

Sporty not Comfortable

As the name implies, the Ducati Sport 48 is an out and out sports bike. It has clip-on handlebars, racing type seat and a downdraught, bell-mouth sports' carburetter. Although the engine is capable of pulling two people, pillion footrests are not fitted. Riding comfort isn't one of the machine's finer points, as the saddle, similar to most Italian makes, is rather narrow and hard. However, apart from this, the riding position is comparable to that of the majority of sports machines.

With a top speed in the region of 46 miles-an-hour, the small full-width hub brakes proved adequate. The back wheel could be locked quite easily, although the front would only attempt to do the same on a damp road.

Lighting was also adequate for performance, although being direct, no parking or stoplights were provided. The horn seemed rather feeble and at low engine revs would hardly do credit to an angry wasp.

One of the problems when riding small capacity machines, is that one often finds that on a standing start from the traffic lights, one usually has a bus or car hammering along close to your back wheel. However, this lack of acceleration doesn't apply to the Ducati. Its three-speed gearbox was fitted with well-chosen ratio cogs and one could keep up with much of the traffic on acceleration.

The twist-grip gearchange operated quite smoothly, although it was possible to slip between the gears into false neutrals if care wasn't taken in selection. Going up through the box it wasn't found necessary to operate the

continued on page 33

PICTURE KEY

1. The neat little clip-on handlebars and controls
2. The rear chain has only a very narrow guard
3. Engine has a very steep downdraught on the carb
4. Ducati has telescopic forks, full-width hubs
5. Handling and comfort is good on all surfaces

ROAD TEST

CYCLE WORLD ROAD TEST

DUCATI DIANA MK 3

OFTEN HAVE WE SAID, "the day of the single is over," and we believe that in time, this statement will become truth. However, for the present, there are single-cylinder engines that do the kind of job that makes our long-term prognosis sound a trifle silly. One of these is the Ducati Diana, in Mk 3 form. It is, single-cylinder engine notwithstanding, the fastest, and nearly the smoothest, standard motorcycle in the 250cc class. Morever, it delivers performance without fussiness, and the example provided us for purposes of testing was returned to its owners only after all excuses for keeping it longer had been exhausted — we liked it that much.

To insure that our test bike was a fair representative of the make and model, we were allowed to choose an example at random, from the distributor's stock-room, and we did the uncrating and assembly ourselves. Nothing special was done with the machine; the hardware that is carried separate from the bike to reduce shipping bulk was bolted on, gas and oil added, and the necessary break-in miles ridden — it was tuned, and then we proceeded with our usual tests.

The performance obtained was quite impressive. The bike was taken to Riverside Raceway for top-speed and handling evaluation, and we were able to get a timed maximum speed on the Ducati. There was an unfavorable wind blowing from time to time, and the speeds varied accordingly, but every run was over the 100-mph mark and the best run, made in still air, was 104.1-mph. With that kind of a maximum, the Ducati Diana is, by a comfortable margin, the fastest 250cc touring bike we have tested, and it ranks right up there with many of the 500 cc machines.

A trip to the drag strip proved that not all of the Ducati's performance was at the top end. Pulling the

standard gearing, which meant that it cleared the end of the quarter while still in 3rd-gear, it delivered a 16.5-second standing-quarter, with a terminal speed of 79.5 mph. Given overall gearing that would allow it to pull maximum revs in 4th-gear at the timing lights, the Ducati Diana would almost surely get up around 85 mph, and that is, if you haven't heard, exceedingly strong — for a 15-incher or a machine of even twice that displacement.

All of this speed comes from an engine that looks good, feels good, sounds good and *is* good. It is a modern design, with unit construction of the crankcase and transmission, and aluminum alloy is used wherever possible. The deep-finned aluminum cylinder has a replaceable cast-iron liner, but the need for replacement should occur only rarely: there are four oversize pistons available, to fit rebores up to .100-inch over standard.

Valve actuation in the Ducati engine is accomplished by a single overhead camshaft and short rockers. Hairpin springs close the valves. The camshaft is driven through bevel gears and a towershaft. It is a superior type of valve-gear layout, in both theory and practice. During our acceleration runs, on one attempt we used a 9500-rpm limit, a full 1100 rpm over the red-line, and touched 10,000 inadvertently on another occasion; there was not a trace of valve float. For racing, it seems likely that the Ducati could be re-cammed to develop maximum power at 9000 rpm, with a safe rpm-limit of 9500. Running the engine faster, with suitable changes to keep it breathing at the higher speeds, would bring a further increase in the already impressive power-output.

The 30 mm Dellorto carburetor unit supplied as standard is a good choice for touring, as it gives a lot of power without sacrificing smoothness, but another Dellorto of somewhat larger size, probably 34 mm, would lift the maximum output. No fiddling with the exhaust plumbing should be necessary: The Ducati Diana is delivered to its buyer with a long, slow-taper, reverse-cone megaphone, in addition to the standard muffler, and this megaphone is perfect for the job. The megaphone was used when we ran performance tests, because it is part of the original equipment. It is necessary to revert to the muffler for street riding, and while that reduces the Ducati's performance somewhat, the reduction is not enough to be of much consequence.

Tech. Ed. Jennings rounds Riverside's turn eight.

Also, a 10,000 rpm tachometer is furnished, one less item the road racing owner won't have to buy. Actually, proper use of the instrument aids even the slowest of transportation riders.

Oil-tightness is not a universal virtue among motorcycles: the opposite is more often the case. But, the Ducati never leaked a drop. Partly, we think, because the engine shares its lower casing with the transmission. This provides a large clearance volume under the piston and thus reduces the "pumping" that occurs as the piston travels up and down. Another item that contributes to oil-tightness is the crankcase breather arrangement: a very large diameter plastic tube leading back to a point above the rear fender. Oil mist blown up into this tube tends to condense on its walls and run back down into the crank/transmission casing, minimizing losses.

This excellent engine is mounted in an equally excellent frame and suspension. The frame is essentially of the tubular-backbone type, but with a front down-tube to secure the engine and tie together the bottom of the structure. With telescopic forks and a swing-arm for the rear wheel, the Ducati's suspension is quite conventional. But, it is not ordinary; it gives a good ride and extremely good handling. The rear suspension's spring/damper units are fitted with folding handles, and these can be pulled out and rotated into any of three positions to accommo-

Frank Scurria, noted AFM rider and employee of ZDS Mtrs. Ducati Distributors in the West, buzzes our traps at over 100 mph.

date varying loads.

The brakes, too, deserve high praise. They are large, and of finned aluminum (with an iron liner) and have more than enough capacity. The front backing plate has a scoop-type inlet and outlet for cooling air. The rear hub carries not only the brake assembly, but a "cush" assembly to absorb shock in the drive.

Handling tests were done at Riverside Raceway, which has just about every kind of corner, and it was while doing this bit of test riding that we *really* began to like the Ducati. The bike's "clip-on" handlebars are too low for touring-type riding, but they are perfect for road-racing. Future models will be delivered with two sets of bars, one a new higher and wider clip-on bar that should be nearly ideal for touring. There is enough wind pressure, when running at high speeds, to take the weight from one's wrists, and the overall positioning of the bars and seat was very good. The seat itself was, unfortunately, too narrow — which seems to be more or less typical of Italian bikes.

Cornering speeds on the Ducati were limited only by the height of the foot-pegs and megaphone — which would "ground" and force the rider to ease-off before the suspension and tires were at the limit. Of course, those pegs are fairly high, and by overall touring standards, the Ducati is cornering very fast indeed when it runs out of "lean." And, under all circumstances, it is steady and controllable, with no bad habits. We all had a whack at "road-racing," and were delighted with its behavior. Not with the handling alone, but with the brakes, and particularly, with the smoothness and willingness of the engine. All theoretical considerations aside, there is a lot to be said for a good, strong single.

The transmission ratios are a bit too widely spaced for road racing, but they are just the thing for touring. The gear change is of the rocker type, and we liked that. Changes, up or down, can be made with a good, solid punch of the foot, and it reduces the chances of missing a shift. The Ducati was quite smooth-shifting, generally, but it was necessary to give the lever a vigorous jab, at times, to get into 4th gear.

The only trace of temperament displayed was an occasional reluctance to start when hot. We think that this failing was almost entirely due to the kick-starter layout, which has the pedal over on the wrong side, for most of us, and had a lever that was simply too short. With the Ducati's 10:1 compression, it takes a lot of pressure to get a rapid run-through, and the short-throw kick lever didn't help things a bit. We did find, though, that it would run-and-bump start very easily every time.

One added reason for the touchy starting was the ignition system, which uses a flywheel magneto in place of the usual Ducati battery and coil. With this setup, unless the engine is run through smartly, there is no spark. Current for the lighting system also came from the flywheel dynamo.

In its appearance, the Ducati Diana is second to none. The overall finish is superb, especially in the care applied in polishing the many aluminum castings. The low bars lend a lot of raciness to its appearance, and so does the big 5-gallon fuel tank, which has knee-notches and a quick-release filler cap that not only look racy but are functional as well. In all, we were absolutely enchanted with the Ducati Diana; we think you will be too. •

DUCATI DIANA MK 3

SPECIFICATIONS

List Price	$689 (FOB L.A.)
Frame Type	tubular, single-loop
Suspension, front	telescopic fork
Suspension, rear	swing arm
Tire size, front	2.75-18
Tire size, rear	2.75-18
Brake lining area, sq. in.	N.A.
Engine type	single cyl, sohc
Bore & Stroke	2.92 x 2.28
Displacement, cu. in.	15.2
Displacement, cu. cent.	248.6
Compression ratio	10.0:1
Bhp @ rpm	30 @ 8300
Carburetion	30mm (1.06") TT Dellorto
Ignition	flywheel magneto
Fuel capacity, gal.	4.5
Oil capacity, pints	5.1
Oil system	wet sump
Starting system	kick, folding crank

POWER TRANSMISSION

Clutch type	multi-disc, wet-plate
Primary drive	helical gear
Final drive	single-row chain
Gear ratio, overall:1	
4th	6.13
3rd	7.45
2nd	10.4
1st	17.4

DIMENSIONS, IN.

Wheelbase	52.0
Saddle height	31.5
Saddle width	4.0
Foot-peg height	9.2
Ground clearance	5.0
Curb weight, lbs.	265

PERFORMANCE

Maximum speed	104
Maximum speed in gears @ 9500 rpm	
4th	113 (not attainable)
3rd	93
2nd	66
1st	39
Mph per 1000 rpm	11.9

SPEEDOMETER ERROR

30 mph	actual
30 mph, actual	no speedometer

ACCELERATION

0-30 mph, sec.	2.9
0-40	4.8
0-50	6.8
0-60	9.2
0-70	12.6
0-80	16.7
0-90	22.2
0-100	33.8
Standing 1/4-mile	16.5
speed reached	79.5

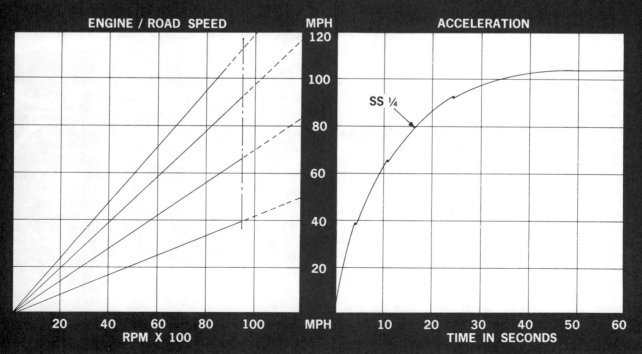

by G. Philipi

Right side of the 1964 Ducati "Mountaineer" reveals the compact lines of the fan-cooled alloy two-stroke engine. Bike has high handlebars, upswept exhaust, 18" wheel in the front and knobby-tired 16" wheel on the rear. Note full-hub front wheel brake.

Testing The Ducati "Mountaineer"

Like lots of other American motorcycle enthusiasts, I ran true to form when I started off on my road riding career. I, like most others, started off with a "big" bike. I must admit I enjoyed the power of big twins and switched from a 74" to a 500 twin, then to a 650 twin and more recently a road version of a 500cc alloy-engined sports racer.

I had an opportunity of talking with Walt von Schonfeld, Public Relations Director for the Berliner Motor Corp., and it was he who spoke of the merits and the fun one could have with a lightweight. I didn't quite believe him, but having the time, visited him at the Norton-Ducati headquarters in Hasbrouck Heights, New Jersey.

On my arrival, I was pleasantly surprised that in spite of being the U.S. Distributors for Norton, Ducati, Sachs and Zundapp machines, the Berliner Motor Corp. also maintains an experimental and research department under the personal direction of Mr. Joseph Berliner. Here "prototypes" of models yet to be produced in quantity are checked and test-driven by some of the nation's top scrambles and enduro riders. It is only after exhaustive tests have been carried out, to the satisfaction of the Berliner technical staff, that the model under test is approved for production. One such test model, a Ducati lightweight, with an alloy engine, three-speed gearbox and 16" knobby tire, had just been returned by a dealer who had covered some 300 miles "cow-trailing" in Sullivan County and another 200 miles on the open highway. This was the bike that Walt wanted me to ride.

The particular Ducati turned over to me for a "test ride" has been dubbed the "Mountaineer" — the engine, a 90cc two-stroke, features a Dell'Orto carburetor with an air-cleaner. The engine has a polished alloy cover (looks as if it were chrome-plated) and is fan-cooled. This means that no matter how slowly the bike is travelling, and no matter what the rpm, the engine always has the proper cooling. The bore is 49mm and stroke 46.1mm.

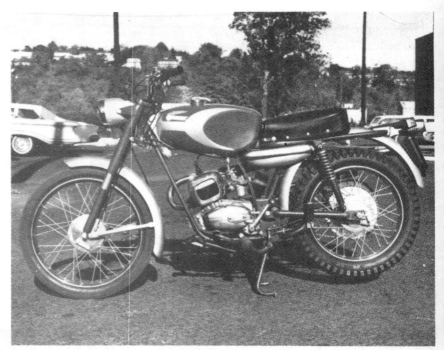

Left side of the 90cc "Mountaineer" shows engine cover orifice to permit warm air to escape, as cool air is forced to circulate about the cylinder as it is forced onto the engine by the fan. Note new "flat-sided" fuel tank, streamlined toolbox, large headlight, taillight, air-cleaner and center stand.

The "Mountaineer" is fitted with telescopic forks, has a robust swinging arm rear suspension and has a pair of tough shock absorbers fitted in the rear. The fuel tank is of entirely new design and has rather "flattened" sides, allowing the rider to hug the tank with his knees when negotiating rough and irregular

To start off, let me say that after Walt "checked me out" on the 1964 Ducati "Mountaineer," I first rode it up and down the macadam roads adjacent to Teterboro Airport, occasionally leaving the paved surface to try the "rough" along the edge of the roads. After being jolted in and out of pot-holes, etc., I

that I may have bitten off just a little more than I could chew! What's more, I was now exhibiting not only the durability of the bike, but also my experience and ability as a scrambles rider!

With a photographer present to record my mechanized gymnastics, I revved the engine in first gear and proceeded to bump down the railroad ties. In many instances, both the front and rear wheel dropped right down to the rough stones between the ties; a wild horse or bucking bronco never gave a rider a rougher trip. The bike had all the stamina to keep on going, but did I? The photographer snapped away and shouted "faster" — now bumping along at approximately ten miles an hour is tough enough, any more throttle would force the front wheel up in the air, and that's exactly what Walt had in mind. He never did tell me that he expected me to do a "wheelie" on the tracks!

DUCATI TEST

Continued from page 27

clutch, and although you could use the same method coming down, it isn't advisable if you hope to obtain a useful life from your engine unit.

One of the beauties of owning a 50 c.c. bike is that you put a gallon of petrol in the tank and don't bother to fill up again for a week or so. The Sport 50 averaged around 105 m.p.g. and that included some fairly hard riding. One unfortunate comment we have to make about the carburation is the lack of air filter/silencer.

The induction roar seemed ten times louder than the exhaust note which was subdued and pleasant under all circumstances. The only way you could keep the carburetter quiet was to use a very gentle hand on the throttle control.

Generally speaking, the Ducati Sport 50 is an economical sports lightweight, which is very reasonably priced and, although it has some utility qualities, is reasonable value for money. It is ideal for the youngster who requires his " first " machine to be cheap, economical and yet sporty. ●

The distances between the ties varies, and here as I travel the bumpy road, the front tire, wheel and fork take a real beating. (so did I!)
(Editor's note: Looks like the front of the exhaust pipe would dent the fender!)

terrain. The handlebars, too, are of special design and have been reinforced with a cross-member for extra rigidity. There's a new style saddle that affords all the comfort one could expect when trailing off the road, and the extra large 61-tooth sprocket that was fitted when I took the bike gave me the feeling that it would climb the side of a building. Berliner Service Technician Mr. Henry Rissman informed me that 13, 14 and 15-tooth countershaft sprockets can be fitted, and a selection of rear sprockets ranging from 36 to 61 teeth offer a range sufficient to satisfy any rider over any type of terrain.

felt that I had the "feel" of this potent little Tiger, and coming back to the Berliner building reported that I had experienced a genuine thrill, my only complaint being that I wish I could have the "Mountaineer" for a longer period of time.

Maybe I boasted a little too much about my ability to ride the "rough stuff," and maybe it was my exuberance over the way the little 90cc Ducati handled, but I was challenged to take the "Mountaineer" over the railroad ties between the rails of the Erie & Lackawanna Railroad. Just as soon as I was on the rough stones and railroad ties, I began to feel

NEW DUCATI FOUR

A RECENT LETTER from the Berliner Motor Corporation informs us that Mr. Joseph Berliner has commissioned Ducati to design, develop and manufacture an all-new, large-displacement touring/sports motorcycle. The machine will be called the D&B V4 (for Ducati and Berliner; and the machine's engine) and it is being produced with funds supplied by Berliner, so that company will of course have a world-wide exclusive distributorship. In the U.S., where it is scheduled to be introduced in mid-1965, the D&B V4 will be known as the Apollo, and the price will be approximately $1500.

Prototypes will be run, in Italy, for a full year before the design is finalized. However, most of the basic features have been decided upon, and the bare specifications for the Apollo are exciting, to say the very least.

The Apollo's engine will be a 90-degree (this configuration has perfect primary balance) V-4, mounted in the frame so that the front cylinders lean forward almost horizontally; the rear cylinders are, of course, tilted back only a few degrees from the vertical. This arrangement gives a good supply of cooling air to both banks of cylinders.

Bore and stroke dimensions are subject to change before the machine reaches production status, but are tentatively 84.5mm x 56mm. Standard and Sport versions of the 1260cc engine will be offered: the former will have a pair of 24mm TT-pattern Dellorto carburetors, 8:1 compression ratio and about 80 bhp at 6000 rpm; the Sport engine will have four 32mm TT Dellortos, 10:1 compression ratio and an estimated 100 bhp at 7000 rpm. This may seem like a lot of power in both instances, and it is, but in a four-cylinder engine, with its relatively huge valve area, the power is not difficult to get. In fact, the 80 bhp engine is being designed as an ultra-reliable, long-distance touring unit, and it is planned that police departments will find it very well suited to their needs. The Sport engine will, obviously, offer absolutely staggering performance at some small penalty in low speed tractability. This is not to say that the Apollo Sport will be a racing motorcycle; it will be a high-speed road machine and will, like the Standard version, be equipped with such civilized accessories as an optional electric starter.

The engine's crankcase doubles as a housing for the 5-speed transmission, and it appears from drawings we were given that a wet-sump lubrication system is employed. In models built for use with a side-car, the transmission will have 4 forward speeds and a reverse.

This very exciting engine/transmission package is hung (literally) in a frame that owes much to the Norton "Featherbed," and Berliner is quick to admit this. The front forks are also Norton-inspired, and we must confess that this is one bit of copying of which we heartily approve. Norton roadholding is legendary.

The Apollo will have a 61" wheelbase, which is a bit long; but not relative to the size of the engine. At present, 16" wheels are specified, but this may be changed before the Apollo reaches production. Indeed, the final product may be quite different in appearance from the artists' rendering. The basic mechanical specifications will be altered very little, but the styling could take a completely different form.

Whatever detail changes are made, the D&B V4 Apollo, when it reaches the public, will almost certainly become *The* prestige bike. It is certain to be impressive in appearance and performance, and the engine will surely offer the smoothest flow of power to be had in any motorcycle today. We have been promised one of the first production models (ah, let it be the 100 bhp Sport model!) and it would be impossible to exaggerate the anticipation with which we look forward to this test. ●

ROAD TESTS OF NEW MODELS

Below: The Daytona has clean, ultra-sporting lines; finish is an attractive blue and silver. Bottom: Power unit has a single overhead camshaft, light-alloy cylinder head and barrel

249 cc DUCATI DAYTONA

ALL the makings of a winner are embodied in this two-fifty Ducati Daytona; sporty, single-overhead camshaft engine, hairline steering and superlative roadholding combined with a handsome appearance make this one of the most attractive lightweights on the road.

If there is one thing that impresses most of all it is the race-bred handling. The Ducati factory have arrived at the ideal combination of steering, suspension and riding position for the most exhilarating of road work. Here is a machine that gives you performance aplenty—plus the confidence to use it to the full.

So good was steering and roadholding that the zestful engine and superb brakes could be used to maximum advantage whenever required. The machine could be cranked over to a really hairy angle without anything grounding; footrests were the first to feel the road—and then very seldom.

Hand in mit with controllability goes the riding position; a bit cut-and-shut, granted, but it has been well-tailored and is just about right, especially for motorway riding. Praise does not extend to the seat which was comfortable only when taken in small doses; after 100-or-so miles one knew it was there!

Of course, the best of handling is no use without an engine to make the frame and suspension work for their living. That of the Daytona has a personality all of its own, besides possessing the qualities traditional to the single-cylinder four-stroke—guts and reliability. It has ample punch for low-speed tick-tocking, negligible vibration throughout the range and, it seemed, unburstability.

Happy cruising speed was between 6,000 and 7,000 on the revmeter (67 to 78 mph) depending on conditions; yet even a severe head wind did not prevent the Ducati pushing along at 65 mph, provided the rider made best use of the low riding position.

Incidentally, the speedometer was between 10 and 15 per cent fast so the revmeter was invaluable in assessing the performance.

Claimed maximum-power revs of 7,500 represents 85 mph in top gear—the speed obtained by stopwatch timing.

Normal changing-up speeds were around 20, 40 and 55 mph. Third, with its 70-plus potential was a very useful gear for slick acceleration when really travelling.

Gear change was positive and smooth although the rocking pedal was pivoted rather far above the footrest for completely easy changes; a lower setting was precluded when it was discovered that the pedal then fouled the exhaust pipe.

For starting, the tickler was never used since it easily flooded the engine, wetting the plug; the air slide was employed only for first starts on cold mornings.

Specification and performance data

ENGINE: Ducati 249 cc (74x57.8mm) overhead-camshaft single. Crankshaft supported in two ball bearings; caged roller big-end bearing. Light-alloy cylinder head and barrel; compression ratio, 8 to 1. Dry-sump lubrication; oil-container capacity 3½ pints.
CARBURETTOR: Dellorto with air filter. Air slide operated by handlebar lever.
IGNITION and LIGHTING: CEV ac generator charging six-volt 13-amp-hour SAFA battery through rectifier; coil ignition with auto-advance. Aprilia 7in-diameter headlamp with pre-focus light unit and 25/25-watt main bulb.
TRANSMISSION: Four-speed gear box integral with engine. Gear ratios: bottom, 18.2 to 1; second, 10.9 to 1; third, 7.8 to 1; top, 6.42 to 1. Multi-plate clutch running in oil. Helical-gear primary drive. Rear chain ½x₃⁄₁₆in with top guard. Engine rpm at 30 mph in top gear, 2,700.
FUEL CAPACITY: 3½ gallons.
TYRES: Pirelli: front, 2.75x18in ribbed; rear, 3.00x18in studded.
BRAKES: Approximately 7in-diameter front, 6½in-diameter rear, both with finger adjusters.
SUSPENSION: Telescopic front fork with hydraulic damping. Pivoted rear fork controlled by spring-and-hydraulic units with three-position hand adjustment for load.
DIMENSIONS: Wheelbase 52in. Ground clearance 5½in. Seat height, 30in. All unladen.
WEIGHT: 280 lb, fully equipped, with approximately one gallon of petrol.
PRICE: £249 0s 6d, including British Purchase Tax. Revmeter £11 10s extra.
ROAD TAX: £2 5s a year.
CONCESSIONAIRES: Ducati Concessionaires, Ltd, 80, Burleigh Road, Stretford, Lancashire.
MEAN MAXIMUM SPEEDS: Bottom, *32 mph; second, *53 mph; third, *74 mph; top, 80 mph. *Valve float occurring.
HIGHEST ONE-WAY SPEED: 85 mph (conditions: light tail wind; 10½-stone rider wearing two-piece leather suit).

MEAN ACCELERATION:

	10-30 mph	20-40 mph	30-50 mph
Bottom	4.4s	—	—
Second	5.2s	5.0s	6.8s
Third	—	6.4s	7.8s
Top	—	7.6s	8.6s

Mean speed at end of quarter-mile from rest: 66 mph.
Mean time to cover standing quarter mile: 18.8s
PETROL CONSUMPTION: at 30 mph, 112 mpg; at 40 mph, 96 mpg; at 50 mph, 72 mpg; at 60 mph, 64 mpg.
BRAKING: From 30 mph to rest, 30ft (surface, dry tarmac).
TURNING CIRCLE: 16ft.
MINIMUM NON-SNATCH SPEED: 17 mph in top gear.
WEIGHT PER CC: 1.127 lb.

Below: Handling is impressive; ideal for twisting roads like this one near Rivington, Lancs. Far right: Neat layout; ammeter is fitted for test purposes only

The engine would start first kick provided it was eased slowly over compression to allow a long swinging prod for maximum effect. Right from cold it would tick-over reliably.

The powerful brakes were smooth in operation yet never threatened to lock the wheels suddenly; the front one was a trifle spongy. The 30ft stopping figure, good as it is, is only part of the story, for these brakes were notably potent also at high speeds.

Unlike most road-test machines, the Daytona was not new or just run in. It had been registered for a year and had been used for demonstration purposes. It was therefore interesting to see how the finish had stood up to time.

All in all it was good, in spite of slight traces of rust on both wheel rims. The blue-and-silver paintwork was still excellent.

The engine was mechanically rather noisy—the large, 300-mile-range tank probably acts as a sounding board for the clattering valve gear—but the exhaust was obtrusive only at revs above, say, 5,000.

Electrics were as reliable as the rest of the bike during the 1,400-mile test. Main beam intensity was so good that it belied its mere 25-watt rating.

Waving goodbye to the Ducati Daytona was a sad occasion. Whether pottering around town or steaming up M1 it was in its element; a machine of which the Italian industry should be proud.

249 cc DUCATI MACH 1

WITHOUT DOUBT, the Italian Ducati Mach 1 is the fastest production two-fifty ever tested by "Motor Cycle." Although it would not pull its 5.39-to-1 top gear to advantage on MIRA's fairly short timing straight, this fifth speed would take the bike over the 100-mph mark on a motorway.

Geared for 115 mph, the bike will fulfil the manufacturer's claim of 106 mph, given neutral conditions and the rider glued to the tank like a pancake on the ceiling. Forgetting top gear, though, the 97 mph best one-way speed in fourth is creditable enough.

In designing the Mach 1, Ducatis have accorded performance the place at the head of the priority table. Road-racing camshaft, 29mm-choke Dellorto carburettor and the 10-to-1 compression ratio serve to make it a machine that is really at home on the track; yet it is pleasantly *au fait* with high-speed, open-road cruising. In town it is definitely *not* at its best.

ROAD TESTS OF NEW MODELS

The riding position is suitable only for a racing crouch—and this can often be embarassing in a busy high street. Obviously, too, the high bottom gear doesn't help matters and a moderate amount of clutch-slipping is necessary to get under way.

The brake pedal is not adjustable and the gear pedal has extensive rejigging precluded by the proximity of the exhaust pipe.

Therefore, unless a rearward riding position is used, it is necessary to lift one's foot off the rests to brake or change gear.

Presumably, he who buys the Mach 1 is not looking for touring comfort. He is after performance. And what a bonus he gets!

There is an old proverb concerning the impossibility of having your cake and eating it; and, it seems, if you want the performance you have to put up with the rest.

Superb acceleration is as much a part of the picture as is top speed. Provided the gears are used to the full, the Mach 1 will show a clean pair of tyres to almost all other two-fifties.

The close-ratio gears make for excellent cruising between 75 and 90 mph—with occasional bursts over the magic ton. The 20-per-cent fast speedometer often registered 110 mph (true 90) with the rider sitting as upright as possible!

In the past few years we have lost old conceptions of relative speeds, capacity for capacity. If we have become

Above: Ultra-sporting engine; the Mach 1 is a Clubman racer in everything but the name. Below: Overall clean, sporty aspect is the intention—colour scheme is a pleasant red and silver

Handling comes second-best only to performance—it inspired full confidence at all times

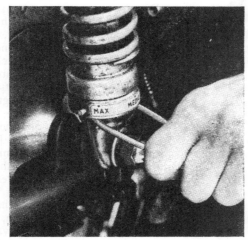

Rear telescopic suspension legs are readily adjustable for loading

SPECIFICATION

ENGINE: Ducati 249 cc (74 x 57.8mm) bevel-driven overhead-camshaft single. Crankshaft supported in two ball bearings; caged roller big-end bearing. Light-alloy cylinder head and barrel; compression ratio, 10 to 1. Dry-sump lubrication; sump capacity, 3¼ pints.
CARBURETTOR: 29mm-choke Dellorto. Air slide operated by handlebar lever.
ELECTRICAL EQUIPMENT: CEV ac generator charging six-volt, 13-amp-hour battery through rectifier; coil ignition with auto-advance. Approximately 6in-diameter headlamp with pre-focus light unit and 25/25-watt main bulb.
TRANSMISSION: Five-speed gear box integral with engine. Gear ratios: bottom, 14.06; second, 9.6; third, 7.49; fourth, 6.11; top, 5.39 to 1. Multi-plate clutch running in oil. Helical-gear primary drive. Rear chain ½ x 5/16in with top guard. Engine rpm at 30 mph in top gear, 2,300.
FUEL CAPACITY: Approximately 3½ gallons.
TYRES: Pirelli 2.50 x 18in ribbed front, 2.75 x 18in studded rear.
BRAKES: Approximately 7½in diameter front, 6½in-diameter rear; both with finger adjusters.
SUSPENSION: Telescopic front fork with hydraulic damping. Pivoted rear fork controlled by spring-and-hydraulic units with hand adjustment for load.
DIMENSIONS: Wheelbase, 51in; seat height, 29in; ground clearance, 6in. All unladen.
WEIGHT: 260 lb, fully equipped and with approximately one gallon of petrol.
PRICE: £269, including British purchase tax.
ROAD TAX: £2 5s a year.
IMPORTERS: Ducati Concessionaires, Ltd, 80, Burleigh Road, Stretford, Lancashire.

PERFORMANCE DATA

MEAN MAXIMUM SPEEDS: Bottom, 45 mph; second, 66 mph; third, 84 mph; fourth, 94 mph; top, 86 mph.
HIGHEST ONE-WAY SPEED: 97 mph in fourth gear (conditions: light tail wind, 10½-stone rider wearing leathers).
MEAN ACCELERATION:

	10-30 mph	20-40 mph	30-50 mph
Bottom	3.1s	3.4s	—
Second	3.6s	3.6s	3.8s
Third	—	4.6s	5.8s
Fourth	—	—	7.2s
Top	—	—	9.8s

Mean speed at end of quarter-mile from rest: 77 mph.
Mean time to cover standing quarter-mile: 17.7s.
PETROL CONSUMPTION: At 30 mph, 144 mpg; at 40 mph, 128 mpg; at 50 mph, 96 mpg; at 60 mph, 62 mpg.
BRAKING: From 30 mph to rest, 30ft (surface, dry tarmac).
TURNING CIRCLE: 14ft 6in.
MINIMUM NON-SNATCH SPEED: 30 mph in top gear.
WEIGHT PER CC: 1.04 lb.

a little blasé, it needs a motor cycle as potent as this Ducati to shake us out of our complacency.

Yet it is rather upsetting to consider that here is a two-fifty that is on the doorstep—if not inside the house—of Mr Average Six-fifty.

The Mach 1 proved a temperamental starter. Liberal flooding of the racing carburettor was necessary most of the time. The ineffective kick starter twined around the footrest in such a way as to be a threat to the starter's shin.

No foolproof starting method was devised. Either it fired first kick or it did not. However, when it proved obstinate, a little athletic sprinting alongside, with bottom or second gear engaged, did the trick.

Once started, it was necessary to blip the throttle to keep up the revs; there was no reliability in the tick-over.

The engine provided above-average clatter, especially from the valve gear; after 3,000 very hard miles, including another road test and a stint of production racing, everything was well and truly loosened up!

Power delivery was slightly harsh, as you would expect from such a high-performance power unit. Moderate vibration was present throughout the rev range, but it never proved annoying.

With so much power on tap, handling has to be beyond fault. It is. Taut and precise steering is combined with rock-steady roadholding. The machine can be leaned over to hairy angles without the fear of anything grounding. Being relatively low, small and light, it can be thrown around to the heart's content.

Handling is probably its star point—in good company with the pokeful little engine.

The brakes are supremely good; they stop the bike with the minimum fuss in the minimum time.

Lighting is definitely below par for a 100-mph sportster. The headlight emits only a yellowish beam that is not satisfactory by any means.

The Ducati Mach 1 is nothing less than a fully-fledged clubman racer with lights added as an afterthought. And, when judged on that basis, it is not far off perfect.

The buyer will have so much to his advantage in performance and handling that the discomfort of riding in traffic will pale, almost, into insignificance.

CYCLE WORLD ROAD TEST

DUCATI DIANA MK III

"ABSOLUTELY ENCHANTED." That was the way we described the feeling of the entire staff when we tested the Ducati Diana Mk III last year, and a test of the recently announced 5-speed version of the same machine has not changed our minds a bit. Rather, we liked the 5-speed Diana even better, and that means we were all falling-down, drooling, in love with the machine. Whatever faults it has, and there are a few, can be traced to the fact that the manufacturers have done everything possible to make it a sports/touring motorcycle with the accent on "sports."

Although largely unchanged from the previous Diana, the newer version is a better bike for having a 5-speed transmission — an unmitigated blessing. Low and high remain the same as before, but there are now 3, instead of 2, ratios between, and the engine can be held somewhat closer to its peaking speed. In the stock Diana, this means that a standing-start 1/4-mile is covered in 16.1-seconds instead of 16.5 (as with the otherwise identical 4-speed Diana). Speed at the quarter-mile's end goes up by about the same percentage: from 79.5 mph to 81.0 mph. Bear in mind that this increase has come without *any* increase of power from, or stress on, the engine.

For the racing fraternity, the 5-speed transmission has even more significance. When dealing with a 4-speed transmission, a tuner must be careful that in boosting the engine's power output, he does not narrow the power range too much. No matter how much peak power an engine may develop, it will be of little value if the transmission has a few, widely-spaced ratios and the rider cannot keep the engine from "falling off the cam." In its present state of tune, the Diana has a lot of power, and it is spread over a wide rpm-range. It was this wide power range that made the 4-speed Diana such a marvelous performer.

However, now that the Diana has 5 closely-staged

ratios, the engine can be squeezed for even more power, even at the expense of narrowing the power range. This means a bigger carburetor, which we have suggested before, and a different exhaust system. The present very slow taper megaphone (supplied in a box; a muffler is fitted on the machine "as delivered") could be changed to one having a somewhat greater angle of divergence. Such a change in megaphones would narrow the power range, it is true; but it would also boost the peak power and the 5-speed transmission will allow the rider to handle a less broad power range very nicely.

Of course, the man who buys a Diana as a fast, lightweight touring machine will not care about any of that racing jazz, but he will be pleased by having a gear for every occasion. Those mountain grades where most small-displacement bikes can't quite pull top gear, and 3rd buzzes the engine too much, are where the 5-speed Diana shines. The tourist will also like the nice, positive feel the transmission has: it "snicks" through the gears in a very satisfactory manner. The one thing *nobody* will like about the transmission is that neutral is all but impossible to select with the bike stopped and the engine running. We soon learned to nip into neutral before coming to a halt, but even then it would sometimes take several tries. This small problem is probably due to having 5 sets of gears stuffed in where there were only 4 previously — which leaves little room for the neutral position.

The other change in the Diana, the handlebars, we are not sure about. The confirmed road rider will like the new western bars better, as they are up where one can reach them without assuming the racing crouch. But, the rider who likes to take Sunday afternoon jaunts up through the mountain roads and make like Phil Read will miss the low, road-racer clip-on bars. Of course, these riders can always order the low bars, and it is probably true that the majority will like those now being supplied on the Diana.

At first, the touring-only rider may find the Diana's engine a bit radical. The engine will run smoothly enough once the machine is underway, but it is a bit too finely-strung to permit much plunking along with the engine at idle. Also, it is something of a grouch about starting. The ignition system gets its current directly from a crankshaft-mounted generator, and a good, rapid run-through is needed to get enough spark for starting. And, too, there is a lot of ignition advance and the engine has a tendency to bite back — hard. After some experimentation, it was discovered that the Diana would start with very little urging if flooded with the float "tickler" and then given a couple of vigorous stabs at the kick lever.

The Diana is equipped, as standard, with a tachometer, and while riding along you will find that this little single winds at an astonishing rate. In normal cruising, the rider will be seeing about 6000 rpm on the tach constantly, and the engine will turn up to 10,000 rpm if he wants to force things. There is a red-line on the tachometer, however (at 8500 rpm), and this should be observed. Maximum power is reached at 8300 rpm, and while twisting the engine higher will get the speed up a little faster, it is an increase that will be paid for sooner or later. The headlamp is the same as fitted on the Ducati Monza, the pure touring 250 Ducati, and contains a speedometer. It is easily removed for racing.

Again, the racing rider will be doing things a bit differently. For racing, the Ducati engine can be modified to peak (with even more than the stock 30 bhp) at 9000, and the rider will probably use 9500 rpm. Obviously, engine life will be shortened, but not by as much as one might think. Even though the Ducati is a "single," and singles are not supposed to be turned too tight, it is a very exceptional single. The bore (2.92") is substantially greater than the stroke (2.28") and having such a short stroke (little more than the Honda 250cc twin) it can crank off a lot of rpm without suffering unduly in the process. Also, the rest of the engine, design and construction, is first-rate. The crankshaft and connecting rod (both very sturdy) run in ball and roller bearings, and are copiously lubricated by oil fed from a gear-type pump. To prevent oiling, the piston is fitted with two oil-scraper rings, in addition to the normal compression rings. The makers have obviously anticipated that buyers do not always observe rev limits because the engine has its valves

operated by short, light rockers from an overhead cam shaft, with racing hairpin springs to close the valves. We tried the Diana at 10,000 rpm and even at that very high speed found no sign of valve float.

The overhead camshaft is driven by spiral-bevel gears through a tower shaft, similar to the arrangement on a Manx Norton. Also similar to the Manx engine is the fact that all of the major engine castings are of aluminum, but where the Manx's cases, etc., are rather rough sand castings, those for the Ducati are cast in permanent molds and are beautifully smooth. Indeed, the Ducati has one of the loveliest (to those who like machinery, and we do) engines being made today. It looks right, and it is right — and it will stay that way. The engine and transmission share the same casing, so there are no external oil lines to leak. The oil supply for both is carried in a sump that is part of the crank/transmission casing.

More bright aluminum is found in the brake drums (which have iron liners, like the cylinder) and in the front forks. The parts are thus lighter, and while the forks may not work any better for having aluminum legs, aluminum brake drums do give better braking. We wish the Diana had aluminum wheel rims, as well, but we suppose that is asking for a bit too much frosting on what is already a marvelous cake.

The Diana is built for touring, and racing, and when one rides it a while it becomes clear where the emphasis has been placed. It is an exceedingly stable machine, and can be dropped over in a turn until things begin to scuff along the pavement without ever becoming unsteady. At the same time, the steering is light and precise. The suspension is properly set-up for fast cornering, which is to say it is rather stiffish, and the combination of stiff spring/damper units and small-section tires make the ride decidedly harsh.

Taken as a racing bike, the Diana 5-speed, like the previous 4-speed, has a great deal of merit. The purchaser can take it out of the box, break it in, fit the megaphone (supplied with the machine), substitute racing handlebars (obtainable from the distributor) and go racing. For a slightly more serious effort, road-racing tires and a fairing are good, of course, but the bike will make a respectable showing and provide its rider with vast entertainment in any case.

As a sports/touring bike it is equally good. The tight suspension, good handling and good performance make it the nearest thing to a good 250cc-class road racer you can ride around the public highways, and it is reliable enough to provide many an afternoon of hard riding.

Strictly on its merits as a touring bike, it does not score too well. It has no battery, and its lights seem to be rather uncertain performers when supplied with electricity only by the generator. Then too, one's posterior region tends to go a trifle dead after an hour or so in jolting contact with that hard saddle, and it's a new, softer seat this year! For touring, the very similar, but more comfortably equipped Ducati Monza (which has a battery, and many other miscellaneous items of touring-type hardware) is a better buy — and it too is now equipped with the 5-speed transmission.

Still, the sporting-type riders around the office swear a mighty oath that the Ducati Diana is the model to have (unless you happen to be one of the mud-plugger set, and want the Scrambler), and they may be right. It certainly is sporting, and beautifully styled and finished as well. It may even be that one of Ducati's Dianas will find a permanent home here at 745 West 3rd St., Long Beach, Calif. •

DUCATI DIANA 5-SPEED

SPECIFICATIONS

List Price	$719
Frame Type	tubular, single-loop
Suspension, front	telescopic fork
Suspension, rear	swing arm
Tire size, front	2.50-18
Tire size, rear	2.75-18
Engine type	single-cyl., sohc
Bore & stroke	2.98 x 2.28
Displacement, cu. in.	15.2
Displacement, cu. cent.	249
Compression ratio	10.0:1
Bhp @ rpm	30 @ 8300
Carburetion	27mm (1.06") TT Dellorto
Ignition	flywheel generator & coil
Fuel capacity, gal.	4.5
Oil capacity, pts.	5.1
Oil System	wet sump
Starting system	kick, folding crank

POWER TRANSMISSION

Clutch Type	multi-disc, wet plate
Primary drive	helical gear
Final drive	single-row chain
Gear ratio, overall: 1	
5th	5.73
4th	6.15
3rd	8.00
2nd	10.3
1st	14.9

DIMENSIONS, IN.

Wheelbase	52.0
Saddle height	29.0
Saddle width	8.5
Foot-peg height	10.0
Ground clearance	6.5
Curb weight, lbs.	268

PERFORMANCE

Practical maximum speed (after ½-mile run)	94
Max. speed in gears @ 9000 rpm	
5th	109
4th	102
3rd	78
2nd	61
1st	42
Mph per 1000 rpm, top gear	12.1

SPEEDOMETER ERROR

30 mph, actual	26.5
50	42.0
70	60.5

ACCELERATION

0-30 mph, sec.	3.0
0-40	4.2
0-50	6.6
0-60	8.5
0-70	11.4
0-80	15.3
0-90	21.3
0-100	29.0
Standing ¼ mile	16.1
speed reached	81

Rear wheel view showing the unusual twin silencers that run off a single exhaust pipe

Beautifully-engineered power unit with angled Dell'Orto carburetter develops 18 b.h.p.

Front drum is 180 mm. and a real stopper—progressive and fade free. Pirelli tyres fitted

Marzocchi shock absorber first-class. Note the eas[y] adjustment and workma[nship]

MOTORCYCLE MECHANICS ROAD TEST No. 87

Vehicle Ducati 200 S.S.   **Price new** £199-0-0
Engine 203 cc. o.h.v. camshaft single
Gearbox 4-speed in unit
Final drive Chain with special cush-drive

GENERAL INFORMATION
- Weight: 245 lbs
- Saddle height: 28½ ins
- Turning circle: 16 ft
- Is toolbox lockable: NO
- Is steering lockable: NO
- Fuel tank capacity: 3⅓ galls
- Reserve capacity: ½ gall
- Oil tank capacity: 3¾ pts
- Gearbox capacity: —
- Fuel specified: Super
- Overall consumption: 69 mpg
- Braking from 30 mph: 26 ft
- Acceleration 0-60 mph: 12 secs

SPEEDS IN GEARS
(chart showing speeds in gears 1-6, mph 10-110)

EQUIPMENT SUPPLIED
STANDARD FITTINGS — Pillion footrests

OPTIONAL EXTRAS — C.R. gears, fibreglass tank, fairing, rev counter, sealed beam headlamp conversion

SPARES PRICES
- Engine gasket set: 19s 6d
- Set valves & guides: 66s 0d
- Piston with rings: 72s 6d
- Set of clutch plates: 74s 0d
- Silencer: 240s complete unit
- Pr Exchange brake shoes: 30s 0d

IT'S getting a bit old hat these day[s to] compare lightweights with much bi[gger] machines and means a lot less tha[n it] did a few years ago. But if ever a [bike] DID merit such comparison it's the 200 c.c. Ducati Super Sports. Because [it] is a big bike in every way but eng[ine] capacity; an overhead camshaft cha[rger] that speeds, steers and stops like a g[ood] big-twin yet still somehow manages [to] return true lightweight running costs.

Frankly, I rarely test bikes of under [250] c.c. capacity because it's just not fai[r to] overload them with my bulk. It cuts do[wn] miles per hour, miles per gallon and al[so] handling characteristics. But with one [ex]ception, the Ducati just didn't m[ind]. Acceleration times were cut only sligh[tly] and the maximum speeds through [the] gears with me in the saddle were ident[ical] to the figures returned by a six st[one] lighter rider.

This single exception came to li[ght] under heavy braking when the front fo[rk] hit the stops with a decided clunk. T[his] was partly due to my weight and par[tly] due to the excellence of the front bra[ke] which was quite outstanding. Braking [dis]tance from a steady 30 m.p.h. was a m[ere] 26 ft. with a beautifully progressive acti[on] which pulled up in a straight line witho[ut] any locked-wheel skittering. This stea[di]ness applied to every facet of the Duc[ati] which made it quite an experience to ri[de].

The handsome engine develops 18 b.h[.p.] at 7,500 r.p.m. and gives all the usual r[ev]ing characteristics of an overhead-ca[m]shaft single with very little poke low do[wn] becoming increasingly dramatic as the re[vs] rise. Compression ratio is a modest 7.8[:1] and the engine is very much over squa[re]. The bore is 67 mm. and stroke 57.8 m[m]. Maximum speeds through the gears we[re] 1st 32 m.p.h.; 2nd 55; 3rd 70 and full no[ise] in top gave a speedo indicated 85 m.p[.h.]

BILL LAWLESS TESTS A NEW 'CAMMY' LIGHTWEIGH[T]
THAT SPEEDS, STEERS AND STOPS LIKE A BIG-TWIN
IT'S THE EXCITING SUPER-SPORTS

ROAD TEST

arside view of engine shows clean lines of the unit. tice the finned sump

Steel rims are standard but alloy ones can be fitted as optional extras

which was just over 83 true speed. The speedo was remarkably accurate throughout the range.

The exhaust system is a sort of siamesed-in-reverse, with the single pipe running into a "Y" branch with two silencers. Why Ducati chose this system I don't know but it is certainly efficient. Back pressure is cut to a minimum while the sound level is well within the law. I never got tired of listening to the exhaust note—a lovely hard "rap".

Starting procedure was a little "iffy" and care had to be taken not to drown the plug, but once away the Ducati warmed up very rapidly and ticked over reliably in a very short time. It was necessary to plug the inclined Dell'Orto carburetter with a piece of rag when the bike was left outside for any length of time.

As with any high-performance engine, running-in procedures had to be followed faithfully and a careful 1,000 miles from new would pay big dividends in engine life. Vic Camp Motorcycles, of Queens Road, Walthamstow, who supplied the road test bike, actually fit a governor which limits throttle openings during the running-in period. When there's a thousand miles on the clock they remove it, service the machine—and away you go. I think this is a very sensible precaution.

The Super Sports seemed to have a built-in safety limit which prevented possible over-revving. In second and third gears it quickly reached its maximum and was reluctant to rev any further. This may have been a characteristic of one particular machine but it was very useful.

Gear selection was first class with positive, foolproof changes which made the in-unit box a delight to use. Top gear, however, was very high and virtually useless under 40 m.p.h. Still, you can't have it all ways and top cog provided an easy-paced touring gear. Top end acceleration in second and third was vivid.

In the suspension department, the Ducati is very conventional and very good. The engine is cradled in a single down-tube frame with a very efficient steering damper and telehydraulic forks which are good enough to fit straight on a racer. The rear swinging arm is damped by two three-way adjustable shock absorbers which feature built-in adjusters. Simply turn a lever on the bottom.

All of which means that the Ducati can be belted into all types of bends as fast as the rider can take them while the bike remains as steady and vice-free as a curate running for vicar. If I owned one of these lightweights I would stiffen the front forks, but this doesn't apply to the normal-sized rider.

Front drum diameter is a massive 180 mm. with a 160 mm. drum on the rear. As I said earlier the bike stops as well as it
—*continued on page* 48

DUCATI 200

Road Impression:

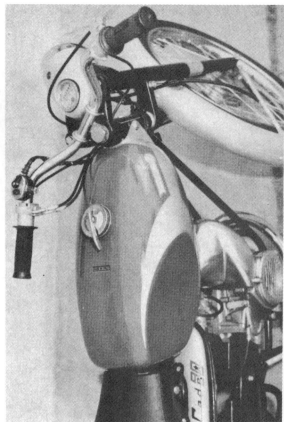

DUCATI CADET 90

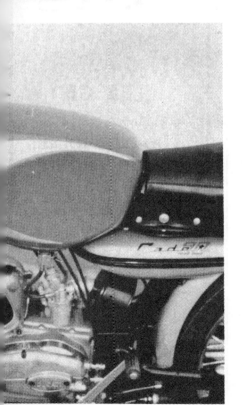

Forced air is one of the more modern technologies being applied to the heating of private residences these days, and it is also one of the latest methods of cooling motorcycle engines. We are not saying it is anything new, even the popularly-used chain saw avails itself of this method of cooling, but it is rarely found on a motorcycle or scooter.

Advantages of the forced air cooling system are simply that the machine need not be moving through the air to cool the engine. This is especially advantageous off-the-road where lower gears are used frequently, and the commensurate higher engine revolutions produce more engine heat. In the fan-cooled engine the higher revolutions actually produce more cooling air, and at a time when it is needed most, incidentally. The advantages of forced air cooling are of course not limited to the trail machine, but this best describes its most important benefits.

Ducati's little 90 Cadet combines this asset, with a number of other attractive reasons for owning one, into a highly desirable machine. The handsome engine is a single-cylinder, two-stroke, of an actual 86.744 cubic centimeters. Horsepower is kept a dark secret, but it is by all standards enough to keep the Cadet in step with its competition. We found fault only with the motor scooter type gear change mechanism on the left handlebar, a minor fault and one that can be corrected by ordering the model with the foot-change system. It is the type of objection we have often found was ours alone so we'll say no more.

Though only three speeds are in the gearbox they seem well enough spaced, and as Ducati traits would dictate, gear changing is pure delight. Finish, paint, polish and the other external methods of judging a machine are up to Ducati's usual standards. Most aluminum parts are polished to a high lustre, painted parts show the careful attention to application, typical of both Ducati and most other machines emanating from industrial Italy.

Top speed is claimed to be approximately 56 miles per hour, and gas consumption around 102 miles per gallon. Both are respectable and highly normal figures for such a machine. Among the many things we liked about the Cadet 90 was the almost totally silent operation, thanks in part to the cast alloy shrouding that surrounds most of the engine, dampening most cylinder noise, and thanks to a nice, big muffler. Large alloy brakes, oil dampened suspension at both ends, dual footpegs, full lighting equipment, chrome luggage rack, chain guard, and tire pump, put the finishing touches to the Ducati Cadet 90. Its price is another thing we think all will like. ●

NEW DUCATI MODELS FROM BERLINER INCLUDE 350cc "SEBRING", 160cc "MONZA JUNIOR" "BIRA" SCOOTER

"Sebring" 350cc

"Monza Junior" 160cc

The new Ducati scooter "Bira" features a healthy 7 hp, 100cc fan-cooled engine.

Latest model of the Ducati line to get the Berliner push is the 1965 Ducati Scooter "Bira", shown here. It has pleasing lines and, since Italian makes of scooters have always had good sales in the United States, certainly the "Bira" built by the famed Ducati firm will find a good market here. It has a 100cc alloy engine, fan-cooled two-stroke, 8.5:1 compression ratio, 7 hp at 5500 rpms. Standard equipment includes 3-speed gearbox built-in unit with engine, comfortable dual seat, large headlight, speedometer, taillight and center stand.

Two new names have been added to the line — the "Monza Junior" and the "Sebring", named after the famous Florida road racing course. "Junior" specifications are as follows: 160cc overhead camshaft alloy engine with a bore of 62mm and stroke of 52mm, compression ratio 7.5:1, 12 hp at 6000 rpms, magneto ignition, comfortable dual seat, chrome luggage rack, adjustable. Shocks, front wheel 2:75 x 16", rear wheel 3:25 x 16", Del'Orto carburetor with air-cleaner, large full-hub front and rear brakes, headlight, speedometer, safety guards, are standard equipment. The maker claims 75 mph and 80 miles to the gallon.

The "Sebring" specifications are as follows: 350cc overhead camshaft alloy engine with a bore of 76mm and stroke of 76mm, 8 to 1 compression ratio, 33 hp at 7000 rpms, battery ignition, large full-hub front and rear brakes, comfortable dual seat, three-way adjustable rear shock absorbers, large headlight with speedometer, five-speed gearbox built in unit with engine, tele-hydraulic front forks, Del'Orto sports carburetor with air-cleaner, steel wheel rims, chrome plated, Tire size: front 2:75 x 18, rear 3:25 x 18", taillight, electric horn, side stand and center stand, chromium-plated exhaust system and muffler, 3½ gallon fuel tank, fully adjustable front brake and clutch levers. A top speed of over 100 mph is claimed by the maker.

DUCATI TEST
continued from page 45

goes. Tyre sizes are 2.75-18 front and 3.00-18 rear. Expect around 70 m.p.g. under normal riding conditions which falls off to 60 ridden really hard.

As far as styling is concerned the little Ducati is a handsome eye-catcher with plenty of polished alloy, chrome, and red and gold paintwork. The only thing that could possibly be faulted in this department was the quality of the chrome especially on the tank. A few outings in wet weather brought a distinct veneer of rust.

The electrics are well thought out. A single switch in the centre of the headlamp operates the lights and a dipswitch and horn button is fitted conveniently on the left handlebar. A stop light works off the rear brake. A six-volt battery mounted between the twin tool boxes and the rear frame member is recharged by an alternator flywheel and rectifier.

To sum up, the Ducati Super Sports is a near-perfect mount for anyone, and just right for the rider who wants an economical lightweight with the emphasis on sporting performance. It demands to be ridden properly and thrives on high revs. It must be one of the brightest newcomers to the 1965 lightweight scene.

DUCATI OWNERS CLUB GB

The club for all Ducati owners.

For full details contact the Membership Secretary

Martin Littlebury,
32, Lawrence Moorings,
Sawbridgeworth,
Herts.
CM21 9PE

Scooter Test: DUCATI PORTABLE

Wheels are borrowed from the Ducati 90cc scooter and measure 8 inches. Star-shaped knobs control folding mechanism.

AFTER SEEING the Ducati Portable and mentioning it briefly (Around The Industry, CW December, 1964), we were of course impressed with the newest and smallest member of the Ducati line of machines. Ducati now comprises a range of motorcycles from 50cc, the portable, to 1260cc, the new Ducati/Berliner four. We might point out that the news of the four was also exclusive in CYCLE WORLD when we announced it in April, 1964.

Being nuts about two-wheeled vehicles, we couldn't wait to get our hands on the Portable. Though the model shown here is still in its development stages, we cannot resist displaying an overt amount of enthusiasm for it. It was designed primarily for the person needing highly portable (i.e., light, folding, and miniature) transportation of a temporary nature. We cannot recommend the Portable as a full-time proposition; it is little more than a mini-bike, but with slightly larger wheels.

We can, though, endorse the little two-stroke as the neatest package of extremely portable transportation we have seen in some time. It folds, by a simple system, into a bundle small enough to fit into the trunk of any automobile, or the back of an airplane. It weighs about 75 pounds; even CYCLE WORLD's Advertising Manager can lift it into the back of his Pontiac, as the photograph shows. On the road it is quite simple to operate, and very versatile.

Cruising speed is around 35 mph, sufficient for most short trips, and mileage is fantastic, as would be expected. The Portable qualifies as a transportation bike, rather than a mini-bike toy, since it has full lighting equipment, three-speed gearbox, luggage rack, fenders, chain guard, and most of the other features found on such machines. The price is around $180.00. Though models that wind up on the show room floors of Ducati dealers may vary slightly, it will matter little which changes are made, the story will be the same. If you, or someone you know, is looking for a bike that will stuff into a small space, be easy to carry, yet when assembled be a usable vehicle rather than a mechanical freak, better take a look at the Ducati Portable. ●

CYCLE WORLD's Jerry Ballard hefts the Portable into his automobile.

Two-wheel brakes are standard, as are side stand and horn.

Presto .... chango! Forks slide back on a rectangular tube.

CYCLE WORLD ROAD TEST

DUCATI 160 MONZA JR.

STRANGE AND WONDERFUL are the ways of the world's motorcycle manufacturers. In general, it is these manufacturers' habit to gradually increase the power and/or displacement of a basic model, so that this year's 250 becomes the 305 or 350 of the next year. However, there are times when these same manufacturers will toss us a curve — as is the case with Ducati's new 160.

Basically, the Ducati 160 is a squeezed-down DM 250, with the bore and stroke dimensions reduced from 75mm x 57.8mm to 61mm x 52mm, dropping the displacement from 249cc to 156cc. This has been done without, as nearly as we can determine, materially altering anything in the engine other than the cylinder and piston, and the positioning of the crankpin in the flywheels. The DM 250's cases are finned around the oil sump and there are fins on the valve-rocker covers; in the 160, these fins are missing. However, there appears to be no appreciable change in the bulk of the engine casting or in anything inside. In fact, the instruction manual is referred to by Ducati as a supplement; most of the specifications, procedures and parts from the DM 250 will apply to the 160.

In a way, it would seem likely that the external differences are there to hide the fact that the 160 is a "squeezed" 250, and that is a bit silly. After all, the disguise isn't all that good, and in any case, the 250 is so good that the worst one could say of the 160 is that it is a trifle heavy for its displacement. And, one can point out that the 160 derivation will surely be the most under-stressed engine to come along in years. Parts that have proved so reliable in even the more highly tuned versions of the DM 250 engine should last indefinitely in the substantially less powerful 160.

Retained with the 250 engine is the Diana's frame. This is a single-loop type frame, with a downtube that terminates at the front of the engine crankcase. The engine is a part of the load-carrying structure. Of course, most of the frame's strength comes from the large-diameter, heavywall "backbone" tube. This gives it enormous torsional strength (essential to good handling) and, unfortunately, uncommon weight. It is this combination of engine and frame, both borrowed from the 250, that gives the Ducati 160 much of its high curb-weight of 247 pounds, which must be something of a record for the class.

The rest of the Ducati 160, too, has the appearance of

having been "borrowed" from other models in the Ducati line, but we are not sure what came from whence. The brakes are typically Ducati, with a 158mm drum up front, and a 136mm drum (with built-in drive cushion) at the rear. These are laced into 16-inch wheels, and we feel that in this, Ducati's engineers made their only real mistake. The front forks are, we think, from the Bronco 125, and these forks, combined with the DM 250 frame, give a steering that does not seem to get along very well with a 16-inch front wheel. The bike feels tippy, and makes its rider yearn for a steering damper — which the bike does not have. On the other hand, when the 160 is leaned into a turn, it settles down and becomes very manageable. Peculiar; but that's the way it is.

It may be that bigger wheels would cure this strange-feeling handling. There is room for them, we think, for the bike's fenders do not fit at all closely to the existing wheels, and give the impression of having been intended for something larger.

In riding, we again found that Italian saddles do not suit American posteriors especially well, and we would have preferred something wider and softer. That was, apart from the handling peculiarity already noted, the only thing found lacking. The ride is good, and the brakes excellent, while the various controls were positioned right and worked right as well. The bike's handlebars are of the high-and-wide type, but for a change, we liked these. The 160 is not really fast enough to make one want to lean into the wind, and that is the primary virtue of the low, flat bars we usually prefer on touring machines.

Starting was very easy. This was not just a matter of being lucky with one example, either, for we also had a Ducati 160 as a runabout at Daytona, and it would start on the first kick every time, too. No fuss, no bother; it simply starts when asked, and that is a nice characteristic for any motorcycle.

As for finish, it was like any Ducati — which is to say, very good indeed. Lots of polished aluminum, and deep-luster chrome, and first-rate paint. There can be no doubt that the people at Ducati know how to make things properly. For the good finish, and agreeable disposition, we liked the 160, and while we do not entirely understand this business of shrinking a 250 to make a 160 (the opposite direction would seem to make more sense), it was a pleasant enough motorcycle. •

DUCATI 160 MONZA JUNIOR
SPECIFICATIONS

List Price	$489.00
Frame Type	tubular, single loop
Suspension, front	telescopic fork
Suspension, rear	swing arm
Tire size, front	2.75-16
Tire size, rear	3.25-16
Engine type	single cylinder, sohc
Bore & stroke	2.42 x 2.07
Displacement, cu. in.	9.5
Displacement, cu. cent.	156
Compression ratio	8.2:1
Bhp @ rpm	not spec.
Carburetion	22mm (.875") Dellorto
Ignition	flywheel generator and coil
Fuel capacity, gal.	3.4
Oil capacity, pts.	5.1
Oil System	wet sump
Starting system	kick folding crank

POWER TRANSMISSION

Clutch Type	multi disc, wet plate
Primary drive	helical gear
Final drive	single row chain
Gear ratio, overall: 1	
4th	8.4
3rd	10.2
2nd	14.2
1st	23.7

DIMENSIONS, INCHES

Wheelbase	52.0
Saddle height	30.0
Saddle width	9.5
Foot-peg height	10.0
Ground clearance	(at stand) 4.5
Curb weight, lbs.	247

PERFORMANCE
ACCELERATION

0-60	19.8
Standing ¼ mile	21.0
Speed reached	61

DOING A DUCATI

Bert Furness of Vic Camp's shows how to strip down all the 200 and 250 cc models

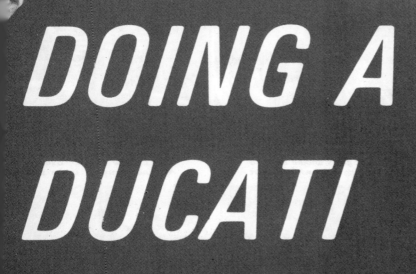

● High-revving, small-capacity engines usually require regular stripdowns without many miles passing under the bike in between.

The rugged Italian Ducatis certainly are the exception to this rule, for although they give high performance, the need for a complete strip is pretty rare.

As good as these motors are, they don't have divine protection and a thoughtless owner can wreck one quickly.

Plenty of oil, changed at regular intervals, is the golden rule for long life. The engine oil must be changed every 1,200 miles, but there is only just over four pints in the sump, so this won't cause any financial heartbreak.

Loss of performance can be traced to maladjusted tappets. The clearances should be inlet 3 thou and exhaust 5 thou on the Mach I, GT and the Monza. The contact-breaker points should be set to 12–15 thou.

DUCATI ENGINE STRIP

1 All electrical connections, the carburettor, exhaust pipe and rear chain must be undone and the selector cover can be removed

2 Before the motor is removed from the frame, the timing marks on the top bevel gears should be lined up by turning motor with kickstart

3 The four head bolts are all th[at] need to be undone to lift off t[he] cylinder head. Usually it [is] necessary to use a soft ma[llet]

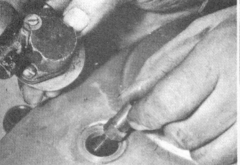

7 On the sporting models, a rev counter is fitted. In this case remove the rev counter drive box before the clutch cover is removed

8 Undo all the necessary screws to release the clutch cover. If it is a tight fit, place a lever underneath and lightly tap cover

9 The clutch plates can be take[n] out by undoing the six screws [on] pressure plate. Striking th[e] screwdriver will ease tight screw[s]

13 The motor should be locked to remove the engine sprocket nut and then the sprocket can be pulled off its keyed shaft

14 The key should be tapped out and then this special tool fitted to the flywheel. The flywheel is pulled off shaft, stator remains

15 The sump filter should be remove[d]. Then the points centre screw an[d] timing cover screws are take[n] out so the timing cover is release[d]

19 Two bolts across the crankcase mouth and all the Allen screws have to be taken out so that the two halves can be taken apart

20 There are shims on the selector drum and on the mainshaft. Label them so that they are put back in the correct place

21 The selector drum, forks and the mainshaft can now be slipped out. The countershaft does not have to be removed from case unless worn

4 With the head off, this sleeve may pop up with the cam drive shaft. If it does, fit it back into place in the lower drive box

5 The barrel can now be pulled up the studs. If this is done with the piston at the bottom of stroke the crankcase mouth is protected

6 Support the piston and drift out the gudgeon pin. The piston may have to be warmed if the pin is a very tight fit in the small end

10 The clutch centre has to be locked in order to undo the centre nut. A special tool is needed to do this effectively without damage

11 The clutch drum will pull off its shaft. Again if it is very tight, it can be eased off with a couple of levers used very carefully

12 The kickstart shaft and return spring are fitted just behind the clutch and can be pulled out by hand. Check splines for any wear

  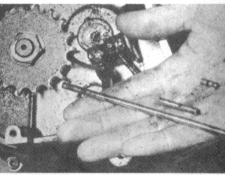

16 A memory aid when reassembling is to position all the timing marks in the timing chest before they are dismantled—there are three

17 Knock down the lock-washer and then undo the left-hand thread nut on mainshaft. Now all the timing gears can be pulled out

18 The clutch push rod assembly will vary slightly from model to model. This Mach I has two rods, two balls and a roller bearing

22 The sludge trap in the flywheels should be undone and cleaned out. This should be a regular job as the build up of sludge is rapid

23 Check the pressure bush for wear on mainshaft. A sloppy fit will lower the oil pressure to big-end and cause the bearing to knock out

24 The oil pump can be checked by priming it and the oilways, and then turning the pump. Pressure should be felt with a finger

CYCLE WORLD ROAD TEST

DUCATI 250 MK III

MOTORCYCLE ROAD TESTS can take some curious turns — especially when being conducted by CYCLE WORLD magazine. In the case of the test performed with this Ducati 250 Mk III, however, we did no more than the character of the motorcycle asked: we ran it in a road race. The reason for this was that while the Mk III has full (well, nearly full) road equipment, it is intended for the ultra-sporting rider, who may just take a notion some fine afternoon to go racing. So, our new staffer, Ivan Wagar, hauled the Mk III off to a race for road-equipped production motorcycles. The bike was in "straight from the crate" form, including ordinary Pirelli tires (which proved to be a mistake) and it started the race with only 96 miles showing on the odometer. A hundred miles later, at the end of the race, the Mk III sailed home in 3rd place and was, if anything, healthier than at the start.

In addition to all that fun, we got a lot of good information from the race experience. For one thing, we learned that the Pirelli tires, while adequate to touring needs, did not have the sort of road-grip required for racing speeds through the turns. Another was that the kick-starter arrangement on the Mk III is ineffective. In these production-bike races, when the flag drops you must run to your machine and start it through whatever normal means apply: kicking, or electric self-start. There is a second flag that gives one permission to get underway free-style (in effect: with a push from helpers). To make starting even a little bit probable, a piece of 2x4 was placed under the center stand (to the amusement of all but the distributor) to permit full stroke of the crank. However, after 30 seconds of frantic effort, Ivan pushed off on the second flag and the engine started immediately.

The starting problem is not typical of all Ducatis; it is created by the footpeg location on this particular model. The Mk III comes with low handlebars (though not, unfortunately, real racing clip-ons) and footpegs mounted high and well back. Thus located, the left-side peg interferes with the kick-start lever's swing, so the starter ratchet, etc., has been jiggled to bring the lever below the peg. This measure, while convenient, leaves one with a very small arc of travel, and it is just not enough to get the engine started unless all conditions are exactly right. We found that to all practical purposes, the only method of starting that could be considered "normal" was the old racing "run-and-bump." And, incidentally, the problem was complicated by the energy-transfer ignition's limited auto-retard which does not spark at all strongly unless the crank is spun briskly, and is likely to produce a whacking backfire when a spark is generated.

Another tricky little feature of this kick-lever/footpeg combination is that unless you fold the peg up and out of the way (the left peg folds; the right is rigid), there is a very high probability that your shin will make a sharp and uncomfortable contact with the peg at the end of the lever's swing. In the end, and after some struggling, we settled on the run-and-bump start as the most satisfactory means of getting the Ducati Mk III underway. We understand that future machines will be supplied with forward mounted pegs, in the more normal manner. For those who prefer the arrangement on the test machine, there will be a kit with the necessary bits; a kit will also be supplied to convert the current model to forward pegs and controls.

Apart from the snappish behavior when starting, we liked the Mk III engine very much. It now has a 29mm TT-type Dellorto racing carburetor and a bigger intake valve, and you get a megaphone as part of the package that gives a noticeable boost in performance. As for the basic design layout: we have liked that since our first meeting. The 250 Ducati is, in many respects, like a little Manx engine. It has a big bore and short stroke, and will run reliably at 9500 to 10,000 rpm in racing form; the stockers will turn that sort of revs but the power fades enough at 9500 to discourage the use of anything higher.

One reason the Ducati 250 engine accepts high speeds without flinching is that it is an overhead camshaft engine. There is just one camshaft, but the rockers used to link cams and valves do not weigh enough to slow things perceptibly. It is fashionable, among Ducati speed tuners, to replace the hairpin-type valve springs with coils, but the low vertical spacing required by the hairpins does give a short, light valve, and the ports can be made a bit more straight-in because these springs are employed. Incidentally, although the usual adjusting-screw arrangement provides for valve-lash settings on touring Ducatis, the Mk III has cup-shims that fit over the ends of the valve stem to do the job. This is weight removed right out at the end of the rocker, where it has maximum effect, and it is another reason for the high-revving capabilities of the Mk III engine. Reliability comes from the exclusive use of large load-capacity ball and roller bearings in the crank assembly.

Tucked away in a back compartment of the crankcase, and sharing the engine's oil supply, is one of the best transmissions being offered in a mass-produced motorcycle today. This is a 5-speed unit, and the ratios are staged like those of a full-fledged racing motorcycle. There is a gear in there for every situation. This box is somewhat sticky about going into neutral at times, but it certainly allows one to make the best possible use of the engine's power range. Fifth gear can best be described as an overdrive when standard gearing is used. However, because of the very small gap between 4th and 5th, we suspect it was intended to be just that. Race preparation for Hanford consisted of putting gas in the tank and checking tire pressures only; as a result top gear could not be used. Rocker-type shift levers are typically Ducati (and typically Italian, for that matter) and on the Mk III, the rocker is remote-mounted with a link leading forward to the transmission.

The course where we ran the Ducati, at Hanford, California, provides a rather severe test of both brakes and

handling, and the Mk III scored well in these areas. Actually, the course was too slippery to permit any ear-'ole style cornering, but it is also quite bumpy (we seem to have caught the course owners somewhere between plowing and planting) and the bumps did not affect the bike unduly. The brakes, which were worked hard, were excellent. Anyone going racing seriously could make them run cooler by opening the "scoops" in the front-brake backing plate. These are cast solid and have some dummy slots in them that admit nothing to the inside of the drum.

The Mk III almost has full road equipment: lights and a muffler; but no horn. Perhaps it is intended that one should shout warnings instead of sounding the non-existent horn. Actually, as there is no battery on this model, it may not be possible to have a horn.

One item in the Ducati Mk III's electrics gave us a few moments of pondering. It is a small switch connected in parallel with the ignition system, and if the taillight burns out, you must reach back on the rear fender and flip this switch to direct ground before the sparks return. Sometimes even when we know, we do not understand.

Whatever else can be said of the Ducati Mk III, it was great fun to ride. We would have preferred genuine clip-on bars to the droopy flat bars fitted, but the existing setup makes it easy to switch to the higher bars most people like. The seat was too hard and narrow, but the riding position was very good if you like the stretched-out semi-racing crouch — and we do. We also liked the big, knee-notched fuel tank, with its quick-release flip-up filler cap. There was a 150 mph speedometer, which gave wildly optimistic readings, and a built-in odometer that also had a tendency to get ahead of itself. The tachometer, which (like the megaphone) comes with the machine but is not installed, was as good as the speedo was bad. It has a large, easily-read dial — marked up to a realistic 10,000 rpm — and will tell you just what the engine is doing at any given moment. •

Editor's footnote: Bob Blair of ZDS motors has just returned from the Ducati factory and tells us that Dr. Montano is going to look into the kick-starter problem personally.

DUCATI MK III

SPECIFICATIONS

List Price	$729 FOB L.A.
Frame Type	tubular, single-loop
Suspension, front	telescopic fork
Suspension, rear	swing arm
Tire size, front	2.50-18
Tire size, rear	2.75-18
Engine type	single-cyl., sohc
Bore & stroke	2.98 x 2.28
Displacement, cu. in.	15.2
Displacement, cu. cent.	249
Compression ratio	10.0:1
Bhp @ rpm	30 @ 8400
Carburetion	29mm(1.14")Dellorto
Ignition	flywheel generator & coil
Fuel capacity, gal.	4.2
Oil capacity, pts.	5.1
Oil system	wet sump
Starting system	kick, folding crank

POWER TRANSMISSION

Clutch Type	multi-disc, wet plate
Primary drive	helical gear
Final drive	single-row chain
Gear ratio, overall:1	
5th	5.40
4th	6.12
3rd	7.50
2nd	9.62
1st	14.1

DIMENSIONS, IN.

Wheelbase	52.0
Saddle height	30.0
Saddle width	9.0
Footpeg height	11.0
Ground clearance (at stand)	4.5
Curb weight, lbs.	247

PERFORMANCE

Practical maximum speed	97
(after ½ mile run)	
Max. speed in gears @ 9000 rpm	
5th	115
4th	104
3rd	83
2nd	65
1st	44
Mph per 1000 rpm, top gear	12.7

SPEEDOMETER ERROR

30 mph	actual 26.0
50	44.0
70	61.0

ACCELERATION

0-30 mph, sec.	2.5
0-40	3.5
0-50	5.5
0-60	7.7
0-70	11.0
0-80	15.0
0-90	21.5
0-100	29.5
Standing ¼ mile	16.3
speed reached	80.5

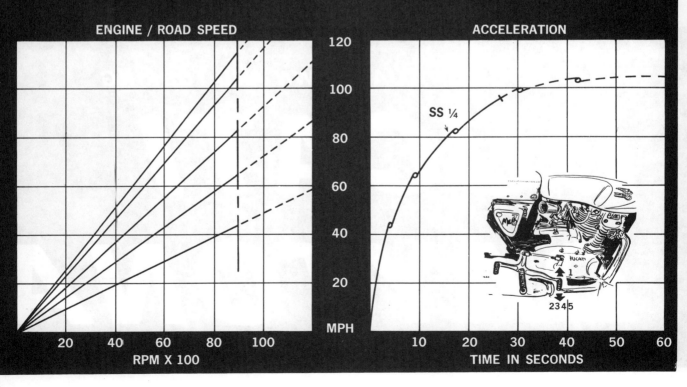

ROAD TEST

FOCAL POINTS

The front brake has air cast into brake plate. front brake cable is lo through the pivot n

MM TESTS THE ITALIAN TOWN AND TOURING
DUCATI DAYTONA

 sual handlebars are a re of the GT. These are ng shape but are fitted same as "clip-on" bars

 A wiring diagram is printed inside the headlamp. Four ceramic fuses are fitted, three in the shell, one on lamp

 The Dell'orto carburettor breathes through a large air cleaner fitted under the cover on the offside of bike

 The best prop stand we've tested—this one folds out of the way while riding but can be operated by seated rider

John Houslander tries out one of the five-in-a-box 250 Italian sportsters

● **The puzzled look on the face of the rider behind you as you change down four times for that tight bend in the road is just one of the pleasures of riding a five-speed Ducati.**

Another is the annoyed look as you out-brake, out-corner and then out-accelerate him. MM picked up one of these fine ohc two-fifty mounts from the Ducati dealer Vic Camp of Walthamstow, for our lesson in five-upmanship.

Basically unaltered, but for the gear ratios, the Ducati GT and Mach I are the logical step from the popular four-speed Daytona. The frame and motors are similar in all models.

The Mach I differs in that it has a more highly tuned engine and a close-ratio box. The GT is considered to be the touring model of the range—which now embraces nine models—but the sporting character is still present.

High-revving motor

The over-square motor (74 mm bore, 58 mm stroke) produces its maximum power at 8,500 rpm. At this figure the claimed output is 23 bhp. The motor is an all-alloy masterpiece of casting and is true unit construction. The head and barrel are alloy with a steel liner in the bore.

The motor boasts several unusual items, the centrifugal oil filter being one of them. There is no filter in the ordinary sense of the word. Instead any impurity is thrown into the sludge trap in the flywheel. The trap has a large capacity and the need for cleaning it only occurs after thousands of miles—usually during a major stripdown.

The engine oil is inserted through a filler in front of the motor. Throughout the entire test, the motor remained dry and oiltight externally.

Starting the bike was not too easy—a largish carburettor, high compression ratio and the kickstart lever on the left-hand side of the bike, all took their toll. When the motor was cold, the float chamber had to be tickled, the air lever closed, ignition turned on and then half a dozen kicks usually got things buzzing. If this failed the plug was probably wetted and it had to be taken out and dried off.

Touring model or not, the GT does have a high-performance motor. The exhaust note was unexpectedly quiet, however. Moving away on a whiff of gas was easy and the seven-plate clutch was first-class. The entire unit is free from mechanical noise which is a pleasant change from the majority of bikes we test. The overhead camshaft driven by a shaft and the geared primary drive both help in this respect.

With a long history of racing successes the Ducati has earned a reputation for ear'oling on the limit. With British tyres this is possible but the Italian ones fitted were a little short of tread on the sides!

The front end seems light but this is soon mastered and the bike can be thrown into corners at high speeds—with nothing touching the deck.

Large diameter full-width brakes are fitted front and rear. They worked without fault and no criticism could be levelled at them.

Italian electrics

The electrical system is certainly worth a mention—the inside of the headlamp looked like Battersea power station! The gear hidden behind the pre-focus unit comprised no less than four fuses, keyed terminal blocks and a printed wiring diagram. Generally the lights were very good but MM testers had a moan about the battery—it is almost impossible to get at without dismantling half the bike.

For quality with a capital Q the Ducati GT is hard to beat—it's no road burner but for everyday motorcycling with a sporty touch the GT has everything.

MOTORCYCLE MECHANICS ROAD TEST No. 107

Vehicle: Ducati G.T. Daytona   Price new £262-10-0
Engine: OHC 248 c.c. Single cylinder
Gearbox: 5-speed unit construction
Final drive: Single row chain

GENERAL INFORMATION
- Weight: 275 lbs
- Saddle height: 31"
- Turning circle: 12'
- Is toolbox lockable: no
- Is steering lockable: no
- Fuel tank capacity: 3¾ galls
- Reserve capacity: ⅓ gal
- Oil tank capacity: 4 pts
- Gearbox capacity: —
- Fuel specified: Super
- Overall consumption: 72 m.p.g.
- Braking from 30 mph: 29'
- Acceleration 0-60 mph: 8.5 secs
  50

EQUIPMENT SUPPLIED
- STANDARD FITTINGS: Prop-stand. Pillion. Foot-rests.
- OPTIONAL EXTRAS: Crash-bar. Rear-rack.

SPARES PRICES
- Engine gasket set: 29s
- Set valves & guides: £2-1-6
- Piston with rings: £4-2-6
- Set of clutch plates: £2-3-9
- Silencer: £5-5-0
- Pr Exchange brake shoes: £1-7-6

...high, apple pie, in the sky, hopes

A DUCATI 250 FOR RACING

ON THE OCCASION OF our road-testing the first 5-speed version of the Ducati 250, we remarked upon the bike's obvious road racing potential. Just fit clip-on handlebars, a megaphone and racing tires, and you can have a low-budget bash at racing. Then, if you like the game, the Ducati always be further modified to make it more competitive. These thoughts came to us in the course of our test, and they must have occurred to many others, for the starting lineup in American road races will always include a flock of Ducatis.

However, it is a long, long jump from a good training machine to a race winner, and while it would seem on the face of it that the Ducati 250 has winning possibilities, not even the Ducati factory has produced one that will run with the Yamaha TD-1B and Harley-Davidson Sprint CR "production-racers." This should not be taken as criticism of the sports-touring Ducati. The 250-class has become ferociously competitive, and the Ducati is not the only otherwise-satisfactory touring 250 that has not been developed sufficiently to become a winner.

Around CYCLE WORLD's offices, the consensus of opinion is that the Ducati does offer the necessary scope for development into a competitive 250. But, opinion was not so optimistic as to obscure the undeniable fact that there is a lot of developing to be done. Therefore, when we actually started our Ducati-for-racing project, the first item on the agenda was a thorough investigation of the motorcycle, just to establish how much could be done with the engine and chassis. Work on the latter was necessarily restricted by the desire, on our part, to build a machine that would be "legal" for AMA Class C competition. Engine modifications, it was decided, should be held to those things available to our readers. We wanted a finished package that anyone, having access to basic machine-shop facilities and with moderate capital, could duplicate.

Our first step was to disassemble everything. Such items as the standard seat and fuel tank were disposed of immediately, being too heavy and otherwise unsuited for a serious racing effort. We would have preferred to discard the frame as well; it is also only slightly lighter than an anvil and a proper double-loop "duplex" frame would offer more rigidity at about half the weight. Unfortunately, the

BY GORDON H. JENNINGS

AMA's technical inspectors would probably not appreciate anything quite so enterprising as a special frame.

Initially, we had thought that some modification of the forks might be necessary, but a close inspection indicated otherwise. The damping characteristics appear to be just what is needed for road racing. The rear suspension's spring/damper units are another matter. People have used the standard units with fair success, but there are replacement ones available that give better results. Ducati makes a racing replacement (these are not always available), and the Italian Ceriani units distributed by Cosmopolitan Motors would do the job. Even so, the convenient selection in spring-rates made us decide upon Girling suspension units. These give virtually no damping on bounce, but a very strong action to restrain rebound, and that is, our experience tells us, exactly what is needed for road racing. We would prefer to fit these dampers with Girling's progressive-rate 60/90 (60 pounds-inch initially, building to 90 pounds-inch at full compression) springs, but these are rarely in stock here in the United States. Lacking those progressive springs, we will start with straight 75 pound-inch springs. Rates up to 90 pound-inches will be tried, and it is anticipated that slightly stiffer springs will be used on fast, relatively smooth courses. Incidentally, the construction of the Ducati forks, which have external springs (we have removed the dust-covers from the forks so these are now exposed), permits experimentation with spring-rates up front. Actually, we do not think this will be necessary but it is nice to know that we have the option of changing the front springs.

About 6 pounds was trimmed from the standard frame by removing the foot-rest brackets and some surplus material around the back of the rear engine/swing-arm mount. More could have been eliminated by drilling holes in everything, but as this would have only amounted to another pound or two, at most, it did not seem to us to be worth the effort. Especially, this swiss-cheese effect looked unattractive to us because of possibly weakening the structure.

NEW MODELS
DUCATI

250cc SCRAMBLER

A new, comfortable saddle tops off the 1966 version of the Ducati 250cc Scrambler. Bigger lights fore and aft and battery ignition will make the Scrambler more useful for highway and city use. Universal tires now fitted are intended to contribute to the dual purpose of this machine. For rough country handling, softer rear shocks and stiffer front fork springs have been incorporated into the latest bikes.

125cc BRONCO

Virtually unchanged for 1966, this campus-transport model has proven dependable in years past. A 125cc overhead valve engine supplies the power and rides two easily. Reportedly, the Bronco gets excellent mileage from regular gas.

100cc MOUNTAINEER

The Mountaineer shares the Cadet's power plant, and also incorporates the four-speed, foot-shift transmission, with the hand-shift now an optional item. It is equipped with off-the-road and highway gearing. High, tucked-in exhaust pipe, knobby tires, reinforced handlebars and plenty of ground clearance will take the Mountaineer over or through just about any terrain.

100cc CADET, 50cc FALCON

One of the few fan-cooled, two-stroke fifties available—a husky, little lightweight that really "earns its keep."

DUCATI Continued from page 63

At present, the CYCLE WORLD Ducati still has its friction-type steering damper. In touring use, this is a good thing, lending stability over humps and bumps in the road surface. Racing is another matter. To get the desired delicacy of control, one must keep the friction damper quite slack; too slack, in fact, to be of much benefit. Where a damper is really needed is in recovering from a slide: the rear wheel steps-out, you correct, and when the bike straightens, its forks will flick right over on opposite lock. This results because "trail" yanks the forks back toward center, after which inertia carries them over near full opposite lock. Following this, they will again be swung back toward center and again inertia will over-do things for the rider. On a good handling bike, these oscillations lose amplitude with each successive cycle, and disappear without doing any damage (usually). Sometimes however, with even the best of motorcycles they will occur with such violence that the rider will be thrown off. The only damper capable of correcting this condition is the hydraulic type, which is velocity-sensitive. Hydraulic dampers have no effect on small, low-speed steering oscillations, but they will prevent the forks from flapping. Before we try any serious racing, the friction damper will be removed, and a hydraulic damper substituted.

A lot of people have tried to make touring brakes do a racing job, and to the best of our knowledge, this has never been entirely successful. So, just to nip-off a budding problem, we ordered a set of Oldani brakes from Italy. These have 200-millimeter (7.88-inch) drums, cast of magnesium alloy with riveted-in iron liners. The front brake has double-leading shoe actuation; the rear, single. A flange is provided for mounting the rear wheel sprocket but it is rigid, instead of the cushion-drive hub of the standard Ducati. To take shock out of the drive, a spring-hub Oldani sprocket is supposed to be used with the Oldani rear hub. We have another idea for the sprocket arrangement which will be explained next month. Cushion-hub Oldani sprockets are rare, and we are trying to use as many readily-available parts as possible.

Clip-on bars, complete with control levers, can be obtained from Berliner Corporation, the Ducati Distributors, and we elected to use these. They are made to fit the bike and you won't find anything better. Fuel tank and fairing both came from Custom Plastics. The use of Goodyear road racing tires should not need explanation. Proper racing tires are absolutely essential, and our experience with the Goodyears indicates that they are at least as good in terms of adhesion as anything available. They are also somewhat expensive, but the wear-rate is so low that when one considers frequency of replacement, the Goodyear tires are a racing bargain.

This brings us to the end of the chassis modifications; at least, until testing indicates what other small changes will be needed. Next month, we will delve into the matter of engine work, which is a great deal more involved.

Cycle Road Test #145

DUCATI 250cc 5-Speed Scrambler

DUCATI 5-SPEED SCRAMBLER

When you buy a Ducati 5-speed Scrambler, you are buying more than just a hot scrambling motorcycle; you are getting a complete competition package that is also rideable on the street. By "complete package," we are referring to the extras that are supplied as part of the purchase. These include three different rear sprockets, an additional countershaft sprocket, complete set of extra cables, a pair of struts to replace shock absorbers for flat-tracking, extra roto caps for valves (various thicknesses), and a set of tools. These add up to a considerably valuable amount dollarwise, but they also reflect the attitude of the Ducati people who have gone all-out to provide a machine that the average enthusiast can afford, and can *win* with on a limited budget.

The power of the overhead-cam Ducati single is well known in racing circles. The 5-speed 250 is rated at 30 horsepower, which means power-plus for hills and acceleration "off the line." Top speed is entirely dependent upon gearing combinations and will vary with each of the four rear sprockets supplied.

For flexibility in every riding situation, a 5-speed is generally superior to 4-speeds. However, the added gear necessitates more rider skill to extract maximum power from the engine at all times.

Throttle response is excellent and this is partially attributable to the gear-driven overhead cam. Fuel reaches the 9.2:1 compression squish chamber via a 27mm Dell'Orto carburetor. Considerable use of the tickler is necessary to properly fill the float chamber prior to starting the machine.

Immediately noticeable when first kicking over this bike is the high compression. This would be less of a bother if the kickstart lever was rubber padded.

Muffler is an option. Bike runs better without it. Note adjustable shocks that require no tools.

Race-bred 4-stroke puts out 30 horsepower. Note protective engine skid-plate. Carburetor is 27mm Dell 'Orto.

Many readers will be interested in the Ducati Scrambler as a street/competition combination. If your plans include mostly street riding with only occasional runs on the dirt, it might be wise to choose another Ducati model more suited to street riding. The Scrambler may be ridden on the street, but there is no speedometer and the bike is delivered without a muffler. Even with the optional muffler, the machine is very loud; but with only a straight pipe, law officers would be able to hear the machine from a considerable distance. Also, long distance rides would prove annoying to the rider. For competition, the standard exhaust pipe is ideal, and the sound is impressive. But a muffler is essential for the street, and the muffler that may be fitted does little to reduce noise, while doing much to hurt performance, since the bike is designed and tuned for no-muffler performance.

Another competition-bred feature of the Ducati 250 Scrambler is the stiff suspension. The '66 model has 25% softer rear suspension than last year's model, but is still quite stiff, even when set on the softest of the three position adjustments. Ideal for the track, but quite stiff for the street. This, however, did not present any problems to our test riders, and some enthusiasts may even prefer the firm ride.

For riders determined to have a competition machine to ride in city traffic, Ducati has provided a fine battery-powered lighting and ignition system, similar to that found on the 160cc Monza Jr. The taillight is identical to the 250 Monza's and is modernistic in appearance. A speedometer may be fitted (extra cost) in the headlight unit.

Improvements for '66 include modified front forks which are stronger and eliminate any possibility of breaking, according to Ducati spokesmen. The hydraulic fork stem, formerly pinned, is now arc-welded to take tremendous impact in jumps. Rear fenders are now secured by strong steel straps. Fittings are now employed to facilitate the attachment of rear passenger pegs.

Seat design in a scrambles machine is important not only for appearance, but for better handling through good rider position. The seat on this Ducati rates as one of the finest we have seen. It is properly contoured and amply padded.

Handlebars are ideally shaped and large padded grips are a big "plus." The tank is handsomely contoured, as are the small, functional fenders. Paint is black and silver, with decorative pin-striping on the tank.

Tires are large Pirelli's: 4.00" x 19" in the rear and 3.50" x 19" in front. Both are "Universals."

SUMMARY:
Our over-all impression of the Ducati 5-speed Scrambler is that it fills a valuable spot in the general motorcycle picture as a machine that is at home on the competition track, but is also suitable for operation around town. It's an extremely rugged bike, both in appearance and performance, and will prove a favorite with 4-stroke lovers who appreciate a highly-tuned, race-bred overhead cam engine.

Part of the package: Sprockets, cables, suspension struts, and tools. No extra charge.

Black rubber cover atop headlight fills spot where speedometer may be fitted as an extra-cost option.

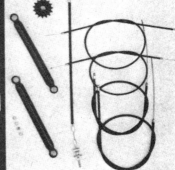

Scooped front brake: a highly desirable feature.

SPECIFICATIONS
DUCATI 250 SCRAMBLER

Engine type	4 cycle OHC
Bore	74mm
Stroke	57.8mm
Horsepower	30 @ 8,700 rpm
Compression ratio	9.2:1
Carburetor	Dell'orto racing 27mm
Fuel tank	3 gal.
Oil tank	2 quarts
Oiling system	Forced by gear-type pump
Ignition	Magneto automatic advance and retard
Front tire	3.00 x 19 motorcross
Rear tire	3.50 x 19 motorcross
Front suspension	Hydraulic telescopic fork
Rear suspension	Adjustable hydraulic damper
Gearbox	5 speed
Front brake	Hand operated, 180mm
Rear brake	Foot operated, 160mm
Weight	242 lbs.
Fuel consumption	68 mpg
Wheel base	52 inches
Saddle height	29½ inches
Colors available	Black frame; black and polished aluminum
Price	$739 FOB Los Angeles or Seattle; $10 less East Coast

A 250 DUCATI FOR RACING
Part Two

WE HAVE REMARKED UPON the Ducati 250's racing potential; previously (before this Ducati racing project), we have not explained why the potential is there. Taken on its design features, the Ducati engine impresses one as being just the thing for racing. The radically "over-square" bore/stroke ratio (1.277:1) would seem to insure large valve sizes and low piston speeds at high revs; the valves are operated from an overhead camshaft; and all of the engine castings are of aluminum alloy. Just as important as these theoretical considerations is the fact that in service, the Ducati 250 has proven to be nearly unbreakable.

Unfortunately, for what we had in mind the standard of reliability established by normal (or even slightly abnormal) service was of limited value. It is axiomatic that maximum power will ultimately come from maximum revs, and our first task was to determine what order of crank speeds the Ducati was likely to withstand. Piston speed is one index of this; a better one is piston acceleration. This is calculated from stroke, connecting rod length, and engine speed. There is no fixed limit for piston acceleration, but considering that the Ducati has a good, forged piston, we tentatively placed our target limit at 125,000 ft/sec^2. That limit is reached at just under 10,000 rpm, but because it is a somewhat elastic limit, the "red-line" our efforts would be directed toward was set at 10,000 rpm. The validity of this pencil-predicted limit is substantiated by the experiences of Frank Scurria, who has raced Ducatis with some success and who found a 10,000 rpm red-line to give reasonable reliability.

Once the upper limit was set, it then became a simple matter to determine the engine speed-range; this being calculated from the ratios provided in the standard Ducati

Only external evidence of engine modifications are oversize carburetor, special intake manifold and Stefa magneto flywheel.

5-speed transmission. To make the rather large jump from 1st to 2nd gear without dropping below the power band, power is required from 6500 rpm. However, on most race courses, one would not use anything under 2nd gear except on the start, so the real power band could be between 8000 and 10,000 rpm — that would cover almost any course and overall gearing conditions we would be likely to encounter.

Knowing how "peaky" power curves become in highly modified engines, we could estimate that with everything tuned for running in a 8000-10,000 rpm range the point of maximum output would be at about 9500 rpm. From that point, it was all fairly easy slide-rule work to arrive at an inlet valve and port size, and a diameter for the carburetor throat.

First to come in for consideration was the intake port diameter. This should be large enough to avoid throttling; yet, small enough to provide the gas velocity needed for good cylinder charging over a fairly wide engine-speed range. Experimental work has shown that gas-flow speeds in the order of 400 feet per second give good results, and a couple of racing engines have used even higher gas speeds. However, some degree of throttling occurs as the flow nears 400 ft/sec, and as we were dealing with a 5-speed transmission, it was not necessary to spread the power band very far. Therefore, we could settle for a narrower range and gain slightly in maximum power. Calculations showed that a port diameter of 1 3/16" would give us a maximum mean gas speed of 380 ft/sec at 10,000 rpm, with a minimum of 307 ft/sec down at 8000 rpm. The lower figure is still high enough to dampen the power-band narrowing effects of a moderately radical racing cam.

At the carburetor, there is no need for having more gas velocity than is necessary for proper air/fuel mixing. In fact, to maintain the high port-area gas velocities out through the carburetor is to incur entirely unnecessary flow losses due to friction. Our ultimate choice of carburetors, an Amal GP5 with 1 3/8" throat, gives a maximum flow velocity of 283 ft/sec; just high enough at 10,000 rpm to create some friction losses. More to the point, the flow at 8000 rpm is 226 ft/sec, which is low enough to make flow restriction minimal and high enough to give completely clean carburetion. Down at the 6500 rpm minimum, gas speed through the carburetor is only 184 ft/sec, and rather borderline for clean running, but acceptable.

Valve size was determined by the interaction of many factors. Contrary to some opinion, biggest is not necessarily best. Insofar as gas-flow is concerned, there is a point of rapidly diminishing returns with increases in valve size, and large valves are a positive embarrassment when dealing with long-overlap valve timing. Moreover, the big valves, which are always added mass in the valve gear, lower the point at which valve-float occurs. Finally, increased valve sizes mean bigger clearance pockets in the piston crown, and make it difficult to get a sufficiently high compression ratio.

By juggling all of these factors around, we settled on an intake valve of 1 5/8" diameter, and a Matchless "twin" intake valve was selected to replace the Ducati valve. As a matter of fact, the latest Ducati 250s have an intake valve only a few "thou" smaller than 1 5/8" and the Ducati valve is made of excellent steel. Unfortunately, the Ducati valve is horribly heavy. The Matchless valve, with its small-diameter stem and deeply tuliped head, is light and has a shape well suited to the port we were planning.

Ah, yes. The intake port. Here was where all the slide-rule work came to a grinding halt and we embarked on a long and frustrating exercise in what is sometimes called "the art of the possible." In other words, making the best of a bad situation. Indeed, we did not fully realize how sticky the situation was until we sawed a cylinder head apart — cutting along the ports. The exhaust port is fine, but the intake port is, according to our Technical Editor, dreadful. While no doubt very easy to cast, being perfectly straight, it directs the gas flow across the valve head, and our tests indicate that only half of the valve circumference is effective to any worthwhile extent. An entirely new port shape would have to be carved into the cylinder head.

In the end, the job was done by milling a new port at 6-degrees downdraft (measured from the plane of the lower cylinder head face), to replace the original 9-degree port. The new port's upper edge starts at the upper edge of the original, out at the port mouth, but because it is slightly larger in diameter, the cut overlaps the bottom of the port. As the cutter moves in, it begins to remove material from the port roof, raising it about .250" above the valve. Hand-finishing (with rotary-files) created a pocket above the valve head to direct flow downward at the point, rather than diagonally across the valve.

All this lifting of the port roof improves flow, but it also brings problems. These originate from the scanty amount of material between the port and the valve-spring/rocker-arm cavity. The port we have described breaks through into this cavity, and the break must be welded. Also, it leaves what may eventually prove to be too little metal around the valve guide, and this is rather poorly supported after the porting work is completed. Frankly, we fear that after some hours of running, the intake guide in our modified cylinder head may come adrift. If it does not, we will have gained a wonderfully smooth port. Should the worst happen, a boss to support the guide will have to be welded into the port.

It has become standard practice, when building a "hot"

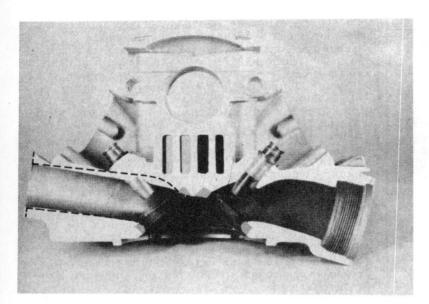

Dotted line on cylinder head cutaway at left shows new port shape. Underside view of head above shows larger port blended into valve seat.

Ducati engine, to replace the stock hairpin springs with coils. Whatever inclination we might have had toward this was removed by the port shape used. To get room for coil springs, it is necessary to cut a relief in the spring-cavity floor at almost precisely the point where we had to weld on aluminum to close a breakthrough in the port roof. Thus, to get a good port shape, one must be prepared to use hairpin springs. As we have found, this is not the handicap it might seem. If the valve gear is light enough, the hairpin springs will do the job, and we had reduced the weight of our intake valve to only 53.9 grams — substantially lighter than a stock Ducati intake valve. (As a matter of interest, we used the original intake valve for an oversize exhaust valve, trimming it to 1 3/8" diameter. It is made of the same steel as Ducati's exhaust valves and was thereby suitable for its change of jobs.)

Machine-shop work, apart from the milling of the intake port, included making special valve-clearance caps, as the Ducati caps do not come in a size thick enough to work with the cam we used, and one had to be made to fit the smaller intake valve stem in any case. The camshaft is a Ducati part, and opens the intake valve 65-degrees before top center; closing it 75-degrees after bottom center. The exhaust valve opens 75-degrees before bottom center and closes 50-degrees after top center. Lift is .380" for the intake valve, and .360" for the exhaust. The valve-clearance problem arises because the extra lift has been obtained by going to a smaller base circle on the cam.

We also machined an intake manifold. This part was machined from the solid, with a flange at one end to bolt against the head, and the other end flanged to match the carburetor. Length is 3" from face to face, as this spacing gave us the correct overall tuned-length for the intake tract. The manifold's bore is tapered from 1 3/8" to 1 3/16" between carburetor and port.

With the valve lifts and diameters we had, the piston-crown clearance pockets were deep enough, but had to be made slightly larger. More depth might have been required, but we used the stock valve seats, and in cutting these for the bigger valves, we also moved the valves deeper into the head. Incidentally, the intake valve seat area was re-cut to give a large radius, or rolled effect, to improve flow when the valve is just off its seat. We might add here, too, that if you attempt to use larger valves than we have recommended, the piston's valve clearance pockets may become so large that there will be too little metal left above the upper ring groove.

Little work is required down below the cylinder head joint. The stock piston was used, as it is a high-quality aluminum-alloy forging, and we did not feel that any of the alternatives offered any particular advantage. A higher piston crown would have been appreciated, as we lost some of the original 10.0:1 compression ratio by sinking the valves in the head, and cutting larger pockets in the piston crown. With those changes, the compression ratio drops to about 9.5:1 and we would prefer 10.5:1. In the future, we may machine the cylinder slightly shorter to move the piston farther into the head and boost the compression ratio up to the desired 10.5:1. A higher compression ratio would yield little gain in power, and with good breathing 10.5:1 will bring the engine near the point of detonation in any case.

One major modification we made that is not, in the strictest sense, entirely necessary, was to install a Swedish-made Stefa magneto. This unit is a standard fitting on the Greeves Challenger, and consists of a small rotor/generating coil/breaker-point assembly that feeds a low-tension output to a separate spark coil. Although primarily intended for two-strokes, it has been used on 4-stroke engines, and its performance in both types of engines has been outstanding.

To install this magneto, we machined a new shaft to replace the one that normally drives the point-breaker cam in the Ducati. The new shaft extends out past the timing case, and the magneto rotor is fixed to a taper at the shaft's end. The magneto's point plate fits into the recess provided for the stock point plate — after the recess is machined to a slightly larger diameter. We added a sleeve just behind the recess, and it holds a ball bearing that supports the shaft. With the added load imposed by

Smiling Frankie Scurria, still recovering from his 1965 Daytona injuries, now serves as CW's builder/tuner/mechanic and worked on much of the Ducati.

the magneto rotor, we did not think that the original brass bushing would be up to the job. The bearing chosen is one with a seal on one side of the races, and this keeps the oil inside the timing case. The most difficult part of this whole conversion was in designing and machining a breaker cam right on the rotor shaft. There is not room to graft-in a breaker cam borrowed from something else.

Two reasons dictated this change. First, we could not use the Ducati magneto because its generating coil does not move with its points, and the points must be set to match the moment of maximum flux in that coil, rather than to whatever moment the engine might prefer for ignition. Thus, it is not possible (or at least not conveniently) to experiment much with spark-lead settings. Most "tuners" work around this by using a constant-loss battery/coil ignition, removing the Ducati generating coils and magnetic flywheel entirely. This works well, but at the expense of carrying a heavy battery — which is also a potential trouble-spot; battery plates sometimes disintegrate due to vibration. Also, a battery/coil system does not match the high-revs performance of the Stefa magneto, nor does it match the magneto's high rate of voltage rise, which makes for great resistance to plug fouling.

The only other modification made below the cylinder head was in the transmission shifter-mechanism. We installed a Sturtevant ratchet-fork, as the older Ducati forks will bend sometimes if the rider attempts to hurry shifts.

Before starting this project, we had the Ducati engine dynamometer-tested, and the results were a bit surprising. With megaphone in place, and after setting the spark and changing jets, to get maximum output, we found ourselves with an honest 22.2 bhp at 8500 rpm. There must be something very special in the air around the Ducati factory's dyno room, for they claim 28 bhp, net, at 8000 rpm (at those revs, our engine delivered 21.3 bhp.) It becomes even more curious when one considers that the Ducati Diana is one of the fastest stock 250s, and virtually all of Ducati's competitors claim much more than 22 bhp. Is it possible that somebody, or several somebodies, has been playing fast and loose with the truth?

There is, at present, no way for us to know how much power we have obtained. The dynamometer facility originally used is currently being renovated, and we do not feel that it would be fair to use other facilities—where we are not certain of accuracy. But, on the basis of such comparative tests as we have made (measuring the CYCLE WORLD-Ducati's performance against the Technical Editor's Yamaha TD-1B), it would appear that we have that 30 bhp and perhaps a bit more. And, we have that wide spread of power. The engine is strong from 6500 rpm right to 10,000 rpm and beyond. It will go all the way to 11,000 rpm without separating, and without a trace of valve float. However, we do not expect that 11,000 rpm could be used very much without suffering a very nasty engine explosion.

Something we did not expect, but are delighted to have, is phenomenally easy starting. Most of our local races are held under FIM rules, with push starts, and if the Ducati does nothing else, it will be first at the first turn. When warm, you take three steps, hit the saddle and drop in the clutch. Invariably, this is followed by an explosion of exhaust noise and a rapidly disappearing motorcycle and rider.

Handling is good, though not outstanding, but the brakes work to perfection. Little pressure at the controls is needed to scrub off speed at a tremendous rate. With this, and the good low-speed pulling qualities of the engine, we expect that our Ducati will be quite a good short-course motorcycle. To make it effective on fast circuits, more sheer horsepower will be required, and we cannot, frankly, afford time for the prolonged development work needed to get that power. The amount of power needed would be in the order of 35-37 bhp, and to get that it will be necessary to direct efforts toward a power peak at probably 10,500 rpm, with the red-line up at 11,000. Reliability at that crank speed will not be good enough using a completely standard crank, rod and piston assembly. It can be done, however, by someone who can afford the time to find ways to keep the engine together while using 11,000 for long periods, and to develop a camshaft to give the horsepower.

At present, we have about 200 man-hours of time invested in the Ducati, and at retail prices (we won't try to kid you; we get a discount on most of this stuff) the CYCLE WORLD-Ducati road racing motorcycle would represent an expenditure of slightly more than $1700. Before the bike reaches winning form (fast enough to compete on an equal basis with the Yamaha TD-1B, Harley-Davidson Sprint CR, or the new 250cc Bultaco TSS) we can envision that price being doubled. Presumably, the Ducati factory could afford this development cost; CYCLE WORLD magazine most definitely cannot afford the sheer number of hours involved — even though some individual enthusiasts can, and probably will.

What we hope we have done is to lay the groundwork for the above-mentioned enthusiasts. All modifications have been purposely held to things the individual can do, and we have given what we consider to be a good basis for further development. With this machine, we have demonstrated that a 1 3/8" GP carburetor works fine; that the stock valve springs do likewise; that you can get a good port carved into the head; and that the engine is quite safe up to 10,000 rpm without bearing, piston or valve-gear problems. From here on, it is every man for himself. It should be mentioned that the wet weight of the bike, complete with fairing, is 233-pounds, or 10-pounds less than the Yamaha TD-1B and probably 40-pounds less than a Sprint CR. Moreover, with a special frame (now possible under the AMA's rules) the Ducati's weight could probably be trimmed another 15-pounds and the handling improved. There is hope, Ducati Lovers; stay with it. ■

CYCLE WORLD-DUCATI ENGINE MODIFICATIONS

Compression ratio	9.5:1
Intake valve diameter	1.587"
Exhaust valve diameter	1.375"
Intake tuned length	13.25" (from valve)
Exhaust tuned length	26.5" (nominal)
Megaphone	31.5" x 3.25"/2.75" (reverse-cone)
Intake port diameter	1.187"
Intake port angle	6-degree downdraft
Exhaust port	standard
Valve timing, intake	65°-75°
Valve timing, exhaust	75°-50°
Ignition	Stefa magneto
Ignition lead	40°
Sturtevant shifter-fork	
Carburetor	Amal GP5
Remote float-chamber	
Main jet	410 (initially)

Spectators at the recent Willow Springs ACA races were treated to a rather strange sight — a single-cylinder four-stroke 250 breaking up the usual parade of Yamaha TD-Bs! Closer inspection revealed it to be the CYCLE WORLD Ducati, ridden by Ralph LeClerq, who has been 175 AFM and ACA champion on a Ducati for two years.

Ralph has purchased CYCLE WORLD's portion of the machine and is carrying on with the development where we left off. Several changes have been made: the Stefa magneto is replaced with a wet cell battery, Ducati ignition points and coil. The system now is the common Ducati total loss racing equipment, completely reliable and very efficient. It is impossible to find fault with this method of obtaining spark, unless one forgets to re-charge the battery. One very definite advantage is that considerable rotating weight has been eliminated, and although the battery weighs more than the magneto, it is "sprung weight" and means nothing in the total picture.

Some changes have been made to the cylinder head. Most drastic was to use longer valves of the same type, and install Webco coil springs, thus eliminating the critical, borderline condition found in the standard hairpin springs at 10,000 rpm. The inlet port had to be built up in the region of the valve guide to give more support to the guide. When port was originally enlarged, machining cut through into the rocker cavity. Two heliarc repair jobs did not cure the tremendous smoking problems, due to oil leaking into the port when the engine had warmed up.

The welds failed for two reasons. First, if the port shape is to remain the same, it is necessary to remove most of the weld from the inside. Not much can be left in the rocker box, as the springs occupy most of the available space. The result is that although a repair has been carried out, there is still precious little to steady the guide, and stresses set up at high engine speeds will simply cause fatigue and lead to further cracking in this thin area.

Valve guide movement was traced quite easily through chatter marks on the valve seat, although the engine had run for only a short period at moderate revs, where valve float would have been impossible. To cure the problem it was necessary to leave considerable meat around the guide and streamline the "lump" as much as possible. In addition, an oversize guide was fitted, as the previously mentioned movement had enlarged the guide hole.

The piston is still the original part giving a compression ratio of slightly less than 10:1. Ralph feels, through experience with the 175 engine, that 11.1 will give better results, and this is the only other actual engine modification he has in mind at this time. Also, it has been found in the past that the oil pump will supply more oil to the valve gear than can drain back to the sump through the standard drain lines when the engine is still cold. This is not a problem on street machines, where the engine is warmed up before high engine speeds are reached. However, it is a factor to be considered on an outright

A 250 DUCATI FOR RACING

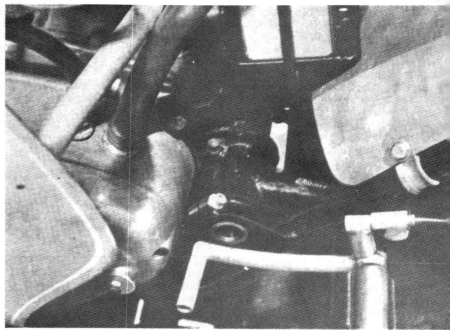

cing engine warming up at four or five ousand rpm, as the over-abundance of l in the rocker box can lead to oil leakg past the valve, until the oil is warm d good drainage begins. To compensate, extra drain line has been installed on e left side of the engine.

Ralph and the CYCLE WORLD expert on road racing feel that good handling can always offset a few horsepower. They feel, and rightly so, that there is no point having extra power if you can't use it. Because of this feeling, the major modifications to the machine have developed.

PART THREE

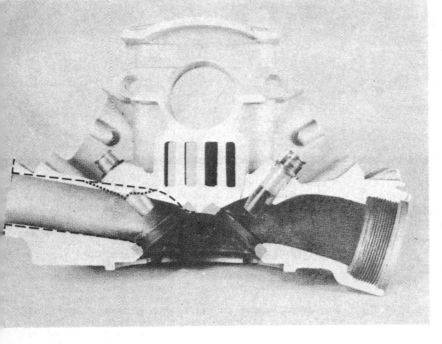

A new swing arm pivot has been made from 4340 chrome moly steel, having a wall thickness of 1/8 inch, and extending to the full width of the rear sub frame down tubes. Short pieces of tubing, with inside diameter to match the new pivot, were welded to sheet metal mounting tabs. A lug was then welded to the tubing, and after saw-cutting through, the lug was drilled and tapped to use a 1/4-20 clamp screw. Then each assembly was pushed on to the protruding ends of the extra-long pivot, which acts as a jig, while the tabs are tack-welded to the rear down tubes.

While this modification is not worthwhile on a street machine, it is an inexpensive, completely satisfactory way to cure many evils, when the extra demands of increased horsepower and bumpy race courses are placed on a touring frame.

To ensure swing arm rigidity, large gussets have been added, which occupy all the space not required by the rear tire when the chain adjusters are at their most forward position. The .080 thick gussets are also made from 4340 and heliarc-welded to the swing arm. While a few pounds have been added to the frame, again it is sprung weight and of little consequence, as the lap times have proven.

Dynamometer testing is planned, but to help evaluate possible performance gain, we took the Ducati to Riverside for some timed runs and were pleasantly surprised to find considerable increases over a standard MKIII. In the standing 1/4 mile we reached a creditable 92.24 mph in 14.29 seconds, and a top speed of 116.12 mph. In standard trim the machine had done 80.5 and 16.3 seconds in the 1/4, while the top speed was 97 mph.

With the available horsepower before frame modifications, the 100 mph "sweeper" was a bit of a handful at Willow Springs, with plenty of jumping about. This means, unless one wants to go on his head, the engine cannot be used to maximum advantage. Yet, with the frame changes, Ralph was able to gain considerable ground on his opponents through this bend.

Of course, nothing is certain in racing. Ralph had fourth place all sewn up after a bad start, even passing and holding off John Buckner. Then, at two-thirds distance, the zipper on his new leathers came undone and a pit stop put him dead last. Lapping at the same speed as the leaders, he pulled back up to fifth place, just behind Buckner. In his quiet, unperturbed way, Ralph is getting a "thumper" into the hornet's nest. ■

Customized DUCATI 200 SUPER SPORT

ROAD TESTS OF NEW MODELS

As sporty as they come, from Goldie silencer to glass-fibre tank. The rear-set gear-change linkage pivots on the folding footrest

A BETTER racer-on-the-road appearance than this machine presents would be hard to imagine. Dropped bars, racing seat, rear-sets and glass-fibre tank; that's the fare from Vic Camp's Ducati showroom in Walthamstow. Vic has taken the already sporty Ducati two-hundred, normal price £210, fitted over £40-worth of alternative equipment (including an excellent Smiths revmeter) and put it on sale for only £239 10s.

The result is a bike that not only looks sporty but feels very much so. The rear-set footrests give a perfect get-down-to-it position, enabling the last ounce of performance to be extracted from this willing little lightweight.

There is no tuning involved in the preparation. It is set up for sporty looks and racy riding position—and that is all.

QUALITIES

Consequently, it loses none of the traditional qualities of the Ducati—the flexible, slogging cammy engine and the taut, precise handling, although comfort is not a strong point.

If one quality has to be selected above all others, then it must be the roadholding. The bike is as sure-wheeled as only a machine with distinguished racing ancestry can be. A real thoroughbred, the Ducati goes just where you put it all the time.

The only debit point in this context is the very firm suspension, even with the rear units set for light loading. It gives a bumpy ride, although this is not detrimental to the roadholding qualities and the racy riding position helps you to make the best of the handling and performance.

The dropped bars are fitted on the standard Super Sport, but there is little doubt that they are more in keeping with the new riding position.

The big question is whether the performance matches up to the ultra-sporting appearance. The answer, bearing in mind that this is only a two-hundred, is a definite yes.

With a best one-way speed of 79 mph (80 is possible in only slightly favourable conditions) and a mean of 77 mph, the Ducati is very fast.

What is more, it is correctly geared. At 77 mph the revmeter reads exactly 7,500 rpm, claimed maximum-power revs.

The relatively close-ratio gear box, with third capable of giving 63 mph, makes it possible to nip along very smartly.

FLEXIBLE

The machine will cruise quite happily on the motorway at speeds approaching 70 mph and the engine is so flexible that it is seldom necessary to change down into third unless a steep hill or strong headwind is encountered.

Like that of most Ducatis, the engine is flexible but the high bottom gear means a fair amount of clutch slipping, especially in heavy traffic.

Consequently, on one or two occasions during the test

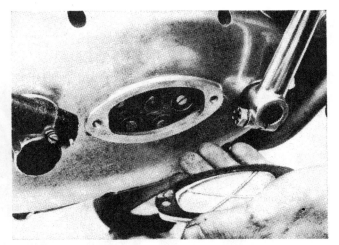

Clutch adjustment and replenishing engine oil is done through this panel on the left side

Neat control layout; the revmeter is particularly attractive, but the ugly number-plate brackets spoil an otherwise good arrangement

Below: Typical Ducati engine—overhead camshaft, all-alloy, deeply finned. Note the drive for the Smiths revmeter and the large-diameter breather tube over the gear box

THE SPORTY BITS

	£ s d
Revmeter kit	12 10 0
Rear-set gear-change and rear-brake levers	6 10 0
Racing seat	4 10 0
3½-gallon glass-fibre fuel tank	8 10 0
Light-alloy rims (pair)	6 0 0
Swept-back exhaust pipe	2 17 6
"Gold Star" pattern silencer	3 5 0
TOTAL, if bought separately	44 2 6

If supplied as original equipment on a new machine, these extras cost £29 10s, included in the "all-in" price of £239 10s.

the clutch became overheated and refused to free completely until it had been allowed to cool down.

Under such conditions, of course, gear changing was difficult.

Normally, however, the change was the sweetest possible. Although not helped by the long, sloppy reverse linkage, the typical smooth, short Ducati change is retained.

The rearward brake pedal, like the gear linkage, was not of the best; it had been shortened to avoid fouling the kick-starter (which it still did, incidentally) and was not so easy to operate as it ought to be.

The brakes, however, are up to the usual high Ducati standard. They stop the bike smoothly and powerfully; **the 35ft from 30 mph performance** is no real indication of the great potency of the brakes at higher speeds.

The engine of the test machine would start first kick every time when warm. Cold starting, however, was less easy, hindered by the fact that the kick-starter fouled the brake pedal on return and had to be freed by hand.

Idling had to be fast— around 2,000 rpm—to be absolutely reliable even when the engine was at its normal working temperature.

On the test model the lighting was below average, being adequate for only about 50 mph at night on unlit roads.

Vic Camp's machine is a bike for the sporty boys right enough. They certainly get what they want—a suitable riding position, really excellent handling and a willing, fast engine. The price is right, too.

Colour scheme is gold and red, with lots of chromium plating

Motor Cycle ROAD TESTS

BOTTOM **SECOND** **THIRD** **TOP**

FUEL CONSUMPTION

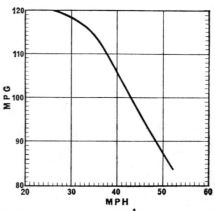

Bottom-, second- and third-gear figures represent maximum power revs, 7,500

SPECIFICATION
ENGINE
 Capacity and type: Ducati 204 cc (67 × 57.8mm) overhead-camshaft single.
 Bearings: Two ball mains; caged roller big end.
 Lubrication: Dry sump; capacity 4 pints.
 Compression ratio: 8.5 to 1.
 Carburettor: Dellorto; air slide operated by handlebar lever.
 Claimed output: 18 bhp at 7,500 rpm.
TRANSMISSION
 Primary: Helical gears.
 Secondary: ½ × 5/16in chain.
 Clutch: Multi-plate.
 Gear ratios: 15.8, 10.87, 8.43 and 6.87 to 1.
 Engine rpm at 30 mph in top gear: 2,900.
ELECTRICAL EQUIPMENT
 Ignition: Six-volt coil.
 Charging: Aprilia ac generator to SAFA 13.5-amp-hour battery through rectifier.
 Headlamp: Aprilia 6in-diameter, with 20/20-watt main bulb.
FUEL CAPACITY: 3½ gallons.
TYRES: Pirelli 2.75 × 18in studded rear; CEAT 2.75 × 18in ribbed front.
BRAKES: Approximately 7in diameter front, 6¼in diameter rear.
SUSPENSION: Telescopic front fork with hydraulic damping. Pivoted rear fork controlled by spring-and-hydraulic units with adjustment for load.
DIMENSIONS: Wheelbase, 52in; ground clearance, 5in; seat height, 28in. All unladen.
WEIGHT: 250 lb, with four pints of oil and one gallon of petrol.
PRICE: £239 10s, including British purchase tax. (In standard trim, without Vic Camp equipment, £210.)
ROAD TAX: £4 a year.
DEALER: Vic Camp Motor Cycles, 131, Queens Road, Walthamstow, London, E17.

PERFORMANCE
(Obtained at the Motor Industry Research Association's Proving Ground, Lindley, Leicestershire).
MEAN MAXIMUM SPEED: 77 mph (11-stone rider in lightweight riding kit; slight following wind).
HIGHEST ONE-WAY SPEED: 79 mph.
BRAKING: From 30 mph to rest on dry tarmac, 35ft.
TURNING CIRCLE: 17ft.
MINIMUM NON-SNATCH SPEED: 20 mph in top gear.
WEIGHT PER CC: 1.23 lb.

ACCELERATION

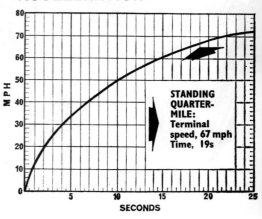

STANDING QUARTER-MILE: Terminal speed, 67 mph Time, 19s

Joe Dunphy tries the ohc
DUCATI
160 Monza Junior

THE Italian Ducati 160 Monza Junior is a machine which will especially appeal to the sporting commuter. It is remarkably economical, covering over 80 miles to a gallon, but at the same time gives an excellent performance for a machine of a mere 156 cc. Propelled by a single-cylinder overhead camshaft engine, with which a four-speed gearbox is in unit, it will show 70 mph on the speedometer.

When I collected this machine from Vic Camp's shop in Walthamstow he informed me that it would have to be run-in before being tested. At first the motor was very tight, but after 100 miles had been covered it became more flexible. The gearbox was also stiff at first.

The Monza Junior looks very sporty with its semi-American-style cowhorn handlebars, and its gay red, aluminium and black colours. An ignition switch is fitted to the top of the headlamp nacelle and is operated by a key, unfortunately the on-off positions are not marked. The speedometer, driven from the front wheel, is also housed in the top of the nacelle. Ball-ended brake and clutch levers and finger-tip cable adjusters are fitted. An air regulating control lever—which I had no occasion to use during the test—is mounted on the right handlebar; and there is a combined dip-switch and horn button on the left.

Parking lights are run off a 6 volt battery and the main headlamp has direct lighting from a flywheel alternator. The weakest feature of this lively Ducati is the lights, it is virtually impossible to use the machine's performance at night if you have to rely solely on this very bad illumination.

The kickstarter is positioned on the left side of the machine as is the cable operated rear brake pedal, and a very useful prop stand, which can be flicked out without any bother, even if wearing cumbersome boots. Gear changes are made with the right foot, one up and three down. A toe and heel pedal is fitted but I preferred just to use the toe part, mainly because it was necessary to lift my foot completely off the pedal after each change when heeling down through the gears.

Close-ups of the single-overhead-camshaft power unit

Although the ratios are suitably spaced, a fifth gear, which could be used as an overdrive when on de-restricted roads or a motorway, would be a welcome innovation. But for general town work the four gears were perfect. The clutch is reasonably light to operate and I experienced no trouble with it.

The engine also gave me no trouble during the period that I used it. Just one or two kicks and it would start every time. Even on some mornings when the temperature was below freezing all I had to do was flood the carburetter, switch on the ignition, kick, and it would burst into life. Maybe roar into life would be a more appropriate way of putting it, for the silencer is not very effective. I didn't really notice this whilst keeping the revs down, during the running-in period, but when I did open her up the difference was remarkable. So much so, that people stopped to watch me pass, especially the odd policeman on duty.

As befits a lightweight with the performance of a much bigger machine, the Monza Junior handles very well, but its Marzocchi telescopic front forks are a bit on the soft side. The rear shock absorbers were right for my 11 stone 4 lb, which was just as well because there are no different settings from which to choose.

The riding position is extremely comfortable and the semi-cowhorn handlebars give full control at the navigation end. I enjoyed mile after mile of riding without any feeling of tiredness whatsoever. The dual seat is firm but comfortable, and the petrol tank fits snugly between the knees.

The 16-in wheels are shod with Pirelli tyres, 2.75-in front and 3.25-in rear. They gave a very good grip in the dry and in the wet.

Capacity of the petrol tank is approximately 2¾ gallons

with about half a gallon reserve. Petrol is fed to the carburetter through two taps with three positions—on, off and reserve.

The engine has a sump holding around 3¼ pints of oil and is pressure lubricated through a gear pump driven by the shaft. The pump takes the oil through a filter from the lowest point of the crankcase, which acts as a reservoir, and forces it through oilways to all parts of the engine requiring lubricant. The oil then returns to the sump by force of gravity. To check the correct oil level, a dip stick has been fitted to the crankcase. But unfortunately it cannot be removed without using a spanner, and the prop stand lying across it makes it difficult to get at.

Accessibility, on the other hand, is excellent when it comes to checking the contact-breaker gap. Only one screw has to be loosened before the cover can be removed, giving easy access to the points.

The full-width brakes are of average effectiveness, but I think different linings could improve them. The front brake is a 6-in single-leading-shoe type, and although the control lever could be pulled back to the handlebar without much effort, the brake appeared to function reasonably well. The rear brake, of 5.3-in diameter, was efficient but I feel that it also could be improved on.

Speeds obtained through the gears were: 22 mph in first; 40 mph in second; 50 mph in third; and, with a slight downhill run, 70 mph in top.

Having covered several hundred miles on this likeable little Ducati, the only really bad point that sticks out in my mind is the very poor lighting. When I mentioned this to Vic Camp, he said that different headlamp units were being tried to overcome this. Also, he intended using thicker oil in the front forks for better damping.

The machine will be available next month and will cost £219, including purchase tax.

CYCLE ROAD TEST

DUCATI 160 MONZA JUNIOR

The clean machine for the clean-cut rider
who likes to indulge in a little sport:
Call it collegiate, and you'll be close.

Just look at it and you know it's supposed to be a man's machine: square-jawed, angular and lean. There's no sissy stuff to get in your way. You start it with a good, strong stomp and you keep it going with a succession of stomps and grabs. The four-stroke single is just boisterous enough to be commanding. And when you ride it, you lean into the low handlebars in a very masculine and sporty way. But it's a sophisticated kind of sporty; it's clean on the verge of suave. Call it collegiate, and you'll be close. The Ducati 160 Monza Junior is made to go with turtleneck sweaters, Cherry Blend tobacco and jewelled fraternity pins.

But it's funny about the man's machine—how it attracts the gentle sex quicker than does English Leather; how, sporting as it is, it doesn't startle. Once Girl has ridden the buddy seat, she'll want to take over the driver's seat. And once she's driven it, she'll want it for her very own. The Ducati goes where she wants it to go and behaves in a manner she appreciates. It doesn't startle. But, on the other hand, and to Girl's great delight, she will (generally) be able to keep up with Boy, even if his is a much larger machine.

The engine is nothing short of aesthetic in design: simple, clean, finished to perfection. And, unlike some engines that look like they are stuffed into or perched precariously on top of a frame, the Ducati engine looks like it belongs. The light alloy cylinder (lined in cast-iron) is closely finned and canted forward ten degrees. Engine capacity is 156 cubic centimeters, overbored at 61mm to a stroke of 52mm. It turns over via a shaft-driven camshaft at a mad pace up to 8000rpm. That, coupled with the sporting 8.2:1 compression ratio and a fair amount of spark advance, should make the Ducati difficult to start. It isn't. You don't even need to touch the handlebar-mounted choke unless it's below freezing. Just tickle the carburetor, find the compression, then swing sharply, with enough pressure to keep your foot firmly on the pedal and you'll get a healthy, first-kick start. If you do it wrong, you'll get a good rap on the instep.

The frame is a single cradle of high-tensile steel tubing; a light, no-nonsense design. That the Monza Junior does not vibrate excessively is partially due to the position of the crankshaft directly on the centerline. It also speaks well of the flywheel balance factor. The ride is firm. Well, the seat is firm, but more than that, the non-adjustable front telescopic and rear swinging arm suspension is set for two-up riding on the street. You feel it when you hit a bump, but the suspension does not top or bottom.

At the same time, the Ducati is lithe. You can whip the front wheel around from lock to lock with your fingertips, if you are so inclined. For one thing, there is no steering damper; for another, the Monza Junior weighs less than 250 pounds. Altogether a very wonder in heavy traffic. The lack of a steering damper had us puzzled (especially since one was mentioned in the manual), but fork wobble was never a problem. At first we thought it might be due again to the well-balanced frame. But as a damper is fitted on the bigger bikes, all of similar design, it must be due to the lightness of the bike and the fact that you don't cruise at speeds all that fast, anyway.

Brake and clutch controls were still stiff after several hundred miles of riding, so it takes some effort to activate them. The plastic horn button and light switches leave a lot to be desired—but then, so do the lights and horn. Parking and stop lights run off the battery; tail and head lamps run off the alternator and consequently, will dim when standing at idle. The horn button is out of reach for the normal hand, not that it matters much when you consider the half-strangled squawk emmitted from that device. Exhaust noise is a far better warning.

Foot-controls are straight out of the fraternity house by-laws on neatness. Right rocker-shift is a heel-toe arrangement that keeps your shoes from getting scuffed. Neutral (*again*) is damn near impossible to find, as you tend to hit second when depressing the toe lever after being in first; and first when depressing the heel lever after being in second, third or fourth. Other than neutral, the gears are positive.

PHOTOGRAPHY: JOHN SENZER

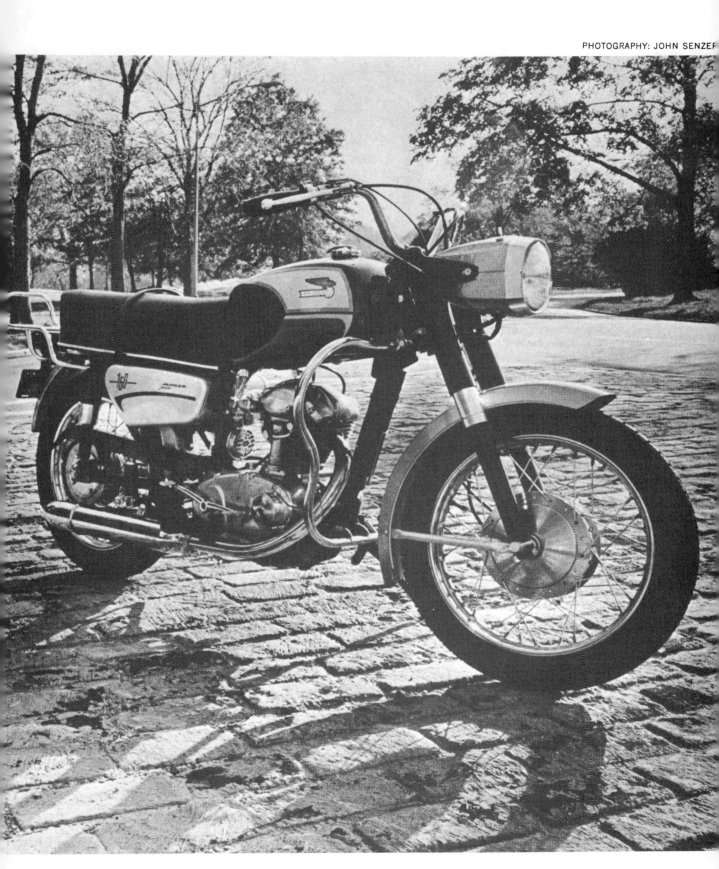

DUCATI 160

The left-hand brake lever is just where it should be when you foot is on the peg. The kick-starter is behind that, and folds neatly out of the way. In front of the brake is a little knob that, when depressed (again with the sole of the shoe), deploys the side stand. One cynical staff member conjured up all sorts of excitement that could result if one were to depress the side stand ejector rather than the brake pedal while in a fast left-hander, but a rider would have to be pretty disorientated to confuse the two.

Single leading shoes are quite adequate for solo riding, or for the speeds that two-up riders are likely to attain. With a heavier rider, the front brake tends to get mushy with repeated use, but for the most part, the brakes are firm —perhaps even sudden—but a lot of that depends on what you are used to.

Into the octagonal headlight nacelle is inset a squarish speedometer/odometer unit and the ignition lock. Ducati expert Malcolm Chu of Berliner Motors, gave us about ten keys to the bike when we went to pick it up for testing. An excellent idea, we thought, considering our previous experience with traditional Italian dinky-keys. They have a nasty habit of popping out of connection, stalling the works and disappearing out of sight. But happily, we report that improvement has been made and we're still using the first key we inserted in the ignition. In fact, there have been times when we couldn't even get it out.

Standard on the bike is a very adequate rear luggage rack, crash bars and tire pump (mounted alongside the front down tube until some squirt decides it will serve his Schwinn well). There is both a side and center stand. Because the bike is so light, the side stand seems superfluous, but we like it anyway and note that it is refreshingly sturdy. Unlike most side stands, it requires fairly level ground—therefore the bike remains a little too upright for comfort—a fact that we discovered in front of New York's famed Plaza Hotel. The ladies in mink thought it rather amusing, but the doorman, directing cabs and limousines through the driveway, was somewhat less than pleased to have a bike lying in the middle of his posh thoroughfare.

Altogether, it's a pleasure to drive the Ducati Monza Junior. You have all the advantages of a tiddler: easy maintenance, sharp response and quick handling. You have the added inducement of a firm sporty ride plus quick acceleration and a respectable top speed (close to 70mph.) On top of all that, the Ducati should be extremely reliable. It can be rather hairy to ride across a windy bridge and very fatiguing on long trips. But what a pleasure to drive from one class to the next. And how impressive parked outside the local hangout.

DUCATI 160 MONZA JUNIOR

Price, suggested retail	$529
Tire, front	2.17 in. x 16 in.
rear	3.25 in. x 16 in.
Engine type	4-stroke single
Bore and stroke	2.40 in. x 2.04 in., 61 mm x 52 mm
Piston displacement	9.519 cu. in., 156cc
Compression ratio	8.2:1
Carburetion	22 mm, Dell'Orto
Ignition	Battery and coil
Fuel capacity	3.43 gal.
Oil capacity	2.2 qts.
Lighting	6 v, 28 watts
Battery	6 v, 7 ah
Wheelbase	52.32 in.
Seat height	28.8 in.
Curb weight	238 lbs.
Instruments	Speedometer, odometer
Top speed	70 mph

The Italians always did have fun with wheels

early Romans are known to have developed the greatest variety of wheeled vehicles for racing, getting ...und and having fun. It's a reputation that Italians still enjoy today. Take their Ducati motorcycles . . . 10 ...ited models that can whip up plenty of excitement. All bikes have the Herculean power for jet-like acceleration . . . the breeding to behave beautifully . . . and the good sense to brake smoothly. So if you'd like to ...ke every day a roamin' holiday, get a Ducati. Your local dealer has 'em all—*lighthearted lightweights* for ... novice and *roadburners* for the seasoned pro. Your choice of OHC and two-cycle engines. You won't find ...er fun-machines anywhere in the world. See the new and exciting 1968 Ducati Models. They're all loaded ... improvements. Your authorized dealer will arrange a test ride, or write for literature and specifications.

250cc Diana Mark III

250cc/350cc Scrambler

160cc/250cc Monza
350cc Sebring

DUCATI
The Thoroughbred of Motorcycles

Berliner Motor Corporation, Hasbrouck Heights, New Jersey

The Sebring, we feel, has the sort of good looks which should appeal to a British public brought up on Tiger 100s and Gold Stars

Well, it's one way to spend a rainy week in November

Testing the 350 Ducati Sebring on the classic Land's End—John O'Groats run

---

DUCATI

The World's Finest and Fastest Motorcycle

All Models from stock

Write for
FREE Catalogue

SPARES : SERVICE
REPAIRS

VIC CAMP MOTORCYCLES

131 QUEENS ROAD,
WALTHAMSTOW, E.17

01-520 2093

SO restricted are we in this country by our road laws that a motorcyclist's only hope for a ride involving unforeseen hazards lies with our unpredictable weather. I had expected, indeed hoped, for some rain . . . maybe an hour or two of torrential road-flooding downpours. I rather enjoy the tussle of keeping a good machine running through a rainstorm while others scurry for shelter or carry on, warm and dry in their heated saloons. It is getting away from the story a mite but . . . a couple of years ago in the Isle of Man during T.T. week when the weather was shocking a couple of friends and I decided to make the best of a wet, stormy day and take a trip over that tarmac track running from the road junction at the top of Injebreck Hill to Ballaugh Bridge. Halfway over the gale-lashed moor we passed a snug little Morris Minor full of snug and derisive youths; and further on we stopped at the Druidale ford in the shelter of an abandoned cottage to enjoy a break. After a minute or two, laughing at the wet and apparently insane motorcyclists, our comfy car friends went hurtling past. They treated the little stream running across the road with all the contempt they could muster and slowed not one jot. In a cloud of spray and steam they climbed the hill on the other side . . . on foot, silently, wet and glum. How do you dry a soaking-wet engine in pouring rain with no waterproofs without becoming very wet? It's wicked, I know, but it was so satisfying. But back to the task in hand. . . .

I had, as I have said, expected some rain—but not the continual downpour encountered for almost the complete journey. Rain began not half an hour after I left home in London for Cornwall and steadily worsened throughout the trip, but the Ducati Sebring, a 350 c.c. tourer supplied by Vic Camp, who incidentally is the new concessionaire for these models, motored along quite happily and so, initially, did I for the old Stormguard coat would keep a man dry under water. We must have looked an odd pair, the gleaming silver and black Italian lightweight, and what must have appeared to be a wet, loosely stuffed bell tent grubily overflowing the seat area. Luckily the Ducati was not one of the sports models for which the factory is so famous: had it been the ride might have had a very different conclusion. As it was, its low bottom gear and touring riding position were just right.

Until Salisbury Plain was reached the rain was merely persistent, but from there on it increased, and so did the wind. However traffic died down measurably so things were pretty good and I was keeping good time, mainly through the Ducati's ability to cruise at whatever throttle opening I wanted. Although its top speed as a tourer was, with all my riding gear,

One of the best points on the machine is that, despite its low weight and small size, riding accommodation is ample for two

cut down to 80 m.p.h., the Sebring would cruise comfortably at that speed for mile after mile; but the 70 m.p.h. speed limit had to be acknowledged so I maintained a reasonably disciplined throttle hand. The weather worsened until, by the time Exeter showed through the driving rain, a full Force 8 gale was blowing straight from the west and into my face. The ride across the Cornish moors was unforgettable. Instead of keeping to the A30, as I had intended, and by-passing Dartmoor, I determined to see the Moor at its wildest, so when the B3213 junction appeared I turned off towards Moretonhampstead and then to Two Bridges and Tavistock. So strong was the wind that I saw little of the landscape. Most of my time was spent in keeping the machine on the road. More than once I was glad of the soft peaty verges when blown across the road and off the tarmac. Much of the road was flooded, often to a depth of three inches, but although I hit some of these rivers at far higher speeds than was sensible, my machine pobbled reliably on. But these high-pressure dousings were causing me discomfort. My body was dry and warm but water had penetrated into my gloves by trickling down my sleeves, and an amazing amount had been forced up between my boot and trouser leg, soaking my feet and legs. Moreover, trickles of ice-cold water had begun to penetrate to the back of my neck despite wool and silk scarves under a securely buttoned storm collar.

Wind-driven rain

On top of Dartmoor I stopped to listen to the wind. No howling gale this, only the ferocious hissing of wind-driven rain on water which, once the engine had stopped, turned the countryside into a very lonely place. I waited only a few minutes, and even then found I had absorbed too much of the atmosphere of the place for my liking. From then on I began looking forward to my hotel with increasing eagerness at every signpost; but for all the anticipated pleasures of a warm bed, through some mule-headed reasoning on my part I decided to see Bodmin Moor as well.

To cut a very long story short, Bodmin was much the same as the other moor, but here a stream had broken up the road surface so badly that this, combined with lumps of peat and debris, caused me to fall off, half ripping the sole from one boot. I decided I would stop at the very next hotel and be damned with Land's End, but even in that frame of mind I found fault with them all. Too big, too small, no garage, any excuse was enough, and so I reached Penzance. Cornish hospitality showed its other side when I stopped at the first hotel I saw but was turned away with the obviously untrue excuse that they were full up. But in the warmth of my next hotel I reasoned that it must have appeared to the first hotel that they were being asked for accommodation by a particularly wet, hairy sack. I had removed my helmet in an attempt to prove that there was a man underneath, but my hair was as wet as my beard....

Drying out

Next day my riding clothes, gloves and helmet especially, were still wet so I stayed in Penzance for the day. The Ducati required no servicing other than a pint of oil in the sump and oiling of the rear chain. It was covered in mud and spray but not a trace of oil stained its engine/gearbox unit or the concrete beneath. Now that, coupled with its vibration-free performance, is what endeared it to me. I would be willing to wager that this machine would travel further with less trouble than the vast majority of motorcycles. An electrical fault did develop later on but that will be mentioned as it arose. A new pair of

---

TOM KIRBY
MAIN ESSEX DISTRIBUTORS
T. W. KIRBY LTD
10 RONEO CORNER,
HORNCHURCH, ESSEX
Tel.: Hornchurch 48785

GEOFF DODKIN
South London's
DUCATI
Agents
UPPER RICHMOND ROAD,
LONDON, S.W.4 Prospect 8779

KENT AREA DISTRIBUTORS
MONTY & WARD MOTORCYCLES
110 HIGH STREET,
EDENBRIDGE, KENT

YOUR MAIN EAST ANGLIAN
DUCATI AGENTS
DELIVERY FROM STOCK
REVETTS (Norwich Rd.) LTD.
53-67 NORWICH ROAD,
IPSWICH.   Tel.: 53726/7

KEYS BROS. for your DUCATI
KEYS BROS.
109 MONTAGUE STREET,
WORTHING
Tel.: 6842

Simple and effective, the three-way load adjuster on the Sebring's spring units requires neither tools nor tears

boots were purchased because of the torn sole, and I was fit for tomorrow's ride. I could ill afford good leather boots while away so wellingtons were my only choice. Words fail me in any attempt to describe my feelings about those boots for my feet experienced misery. If my feet were warm enough to feel comfortable then they perspired, quickly turning the boots into portable foot baths.

Wednesday morning was fine and sunny, so dry and warm I took the coast road to Land's End. R.A.C. Patrolman Weeps signed a postcard purchased from the tourist shop, so that I could prove my story(!), and I left for John O'Groats at 11.45 a.m.

My plan was to travel to Exeter, then start turning up country to Taunton, Bath, Swindon, Oxford, Northampton, and finally arrive at Stamford on the A1, by which time it should be dark, enabling me to travel through to Perth, before having to consult the map again, then through Inverness to the top of Scotland.

The ride to Exeter was almost perfect. It was sunny, and a strong west wind was blowing me along on a barely open

All the tourists had gone home . . . So the road was mine

throttle, and a faint nip in the air added spice to the ride. Not a trace of the storm remained. For the first time I was able to appreciate the Ducati's high-speed handling and braking. All the tourists had gone home and I could see round many of the bends, so the road was mine. We flew along with never a wriggle or twitch even when I was a little over-enthusiastic in cornering and braking simultaneously on a number of bends. Nothing unpredictable happened, so it was obviously all part and parcel of the Sebring riding technique. On we went, in a manner most undignified for a touring motorcycle but eminently suitable for covering many miles in a short time. It struck me after a little while, though, that here was an example of the old story suggesting that it is possible to average high speeds not on the throttle but on the brakes. That was just what was happening. I was approaching corners at higher speeds than I was accustomed to and leaving braking till quite late and in perfect safety, even to the point of braking while cornering. No doubt the modest power output of the engine contributed much to this feeling. I found the rear brake exceptionally good; my usual reaction is to leave the rear well alone for most serious braking and to use it as no more than a steadying device, but the one on this machine was very good, requiring pretty hard pedal pressure to lock it but being powerful and responsive. The front brake, however, was in a class of its own. Like the Triumph 8in front brake, it is a s.l.s. device but had none of the fierceness of so many 2l.s. units. It would pull the machine up with a smooth powerful grip from the highest speeds. It is as good to look at as it was to use, moreover . . . so typical of Italian alloy castings. The brake holds the speedometer drive also; sensible idea this, it should contribute to accurate speedometer readings as the cable is short, free from engine oil, and rev surges. But as seems usual on Ducatis the speedometer itself was a poor instrument. It is completely undamped, so that road shocks and engine vibration had as much effect upon the position of the needle as did the cable revolutions; indeed, while trying for the quarter-mile sprint time it was, according to the speedometer, possible to obtain 70 m.p.h. and no more in every gear save bottom! The position at night is even worse as the face is illuminated from a slot in the front of the instrument that tends to dazzle, and it lies too flat in the headlamp for the figures on the nearer edge of the dial to be read comfortably.

Making good time

At Exeter I stopped to fill the tank. Still in sunshine I pushed on, making good time; it had taken only an hour and 50 minutes from Land's End, and the thought passed through my mind that at this rate I might well reach Stamford in daylight and the Highlands in darkness—something I did not want as it was not to be a record-breaking trip at all and, more important, the Scottish Highlands are far too good to pass through in darkness. So I looked forward to a good high tea while awaiting nightfall at my junction with the A1. But I had not calculated for the average English town council's attempts to keep their town secret from all passing travellers in a maze of badly signposted ring road systems that very often blandly countersign the still existing, old and true, road classification numbers. The game is enlivened by the refusal of so many councils to divulge the name of their village or town. Consequently it is quite feasible for a rider to find that he can be looking for a road junction that common sense tells him will be on the opposite side of the town! It's not so bad in the country. The worst places are mainly the suburban areas of large cities where villages begin blending with characterless "districts." Generally, though, things were going much as planned. I was warm, the roads were still reasonably empty, and the Ducati was banging along with a reassuring regularity.

The north of the Western counties is not well known to me but after my ride through the Glastonbury country I promised myself a holiday there someday. It was so characteristic of how I imagine the English countryside must have been in the days before mechanization spread to farming.

After riding for so long in a sunny West country atmosphere I had begun to feel as though I really was on holiday, just

uring, then suddenly Oxford turned up and it was back to reality, and it had started to rain. Remembering my soaking on the way down, at the next petrol stop I fastened every button and popper and pulled every strap so tight that I was barely able to swing my leg over the saddle, much to the consternation of the pump attendant who tried to help, thinking I had an injured leg from an old tumble; he was an ex-rider himself, he explained, so he understood. The ride to Stamford was uneventful, and by some hard riding and a straight road I was able to make up an hour's time lost in my "holiday" way down the road. By the time Stamford arrived it was dark and I had the feeling that the first leg of the journey was over, but a glance at the map over my meal soon corrected any thoughts of a "quick bash up the road to Inverness." Scotland is only a few miles short of England's north to south length.

Social unacceptability

A disadvantage with motorcycling lies in the social unacceptability of one's clothing; scarcely surprising, as it gathers all the filth that on any other vehicle is regularly washed off. Now I dislike eating in transport cafés which would, I am sure, serve the abominable snowman without questioning his place in the queue for chips. To me a meal is one of the highlights of my day and deserves the careful preparation and comfortable surroundings worthy of such an event. So I searched out such a pub in Stamford. It meant removing my motorcycle clothing before entering the place, but it was well worth the effort. I stayed an hour enjoying a good roast dinner by an open fire (my apologies to Scene's café society). The inner glow from my meal did not last long. By now the rain was steady. Luckily I had once more bound myself hand and foot, with not an opening anywhere, I thought, so now on the featureless A1. I just sat and thought and watched the rain whirl and bounce in silver splinters along the road ahead in the headlamp beam. I had time to think of the bike; its saddle and handlebars, its exhaust note, lighting, controls, fuel and oil reservoirs and consumption—everything, in fact, to do with riding it.

As with all Italian seats, this one felt too hard initially but over a period of more than 2,000 miles I never once felt uncomfortable. It gave a similar feeling to that of the old sprung saddles; and never, unlike most modern over-soft seats, allowed the weight of the rider to crush through to the seat pan, but after a few hundred miles the saddle strap turned from a useful accessory and light backstop into an instrument of misery and discomfort. It was removed. I also blessed my own handlebars, borrowed from my old MSS and fitted in place of the smart little sporty Ducati bars—ideal for the modern image, with a slim, racey look, but not quite the thing for long rides. Our apologies to Ducati, but the MSS bars are shown in the photographs.

The exhaust note was too loud and too penetrating in all performance ranges, and that is about all that can be said about it; it just should have been quieter. All the controls were light, a pleasure to use, the clutch especially. To my mind, it would be hard to find a better one. Starting at the lever and working down, the first impressions are of quality and common sense, because the lever is an all alloy casting pulling a heavy cable sliding through an equally heavy outer sleeve. The entire set up is quite capable of handling a clutch much stronger than the one fitted. The clutch itself is of normal multi-plate type but with the great advantage of running in oil. Consequently it was, as far as I could tell, impossible to cause it to malfunction, despite my slipping it for minutes on end. The oil it runs in, incidentally, lubricates the engine and gearbox also; that is a marvellous idea . . . no messy external oil pipes, no oil tank, no separate fillers for the gearbox or primary drive, and far fewer potential leaking joints.

The engine and headlight suddenly cut. Silence, and darkness. I pulled on to the soft shoulder, and tested the battery. It was full of sparks, so I tested the fuses, and they were all right, and so, physically, was the wiring harness; the only items left were the main switch and the control box, a weird looking thing I dared not think about too much. Everything appeared in order in the headlamp switch but it was all very dead, so

One of the most effective standard brakes we have come across on a road-going machine. The surprisingly soft fork action was most effective

after a lot of jiggling about I traced the main input point and bypassed the switch by wiring it straight to the headlamp switch. The headlamp remained as black as the night. Only the main beam and ignition refused to work, everything else was as it should be since I had rechecked all connections and blindly twitched a desperate finger among the switch wiring maze; this, I suspect, rather than my checking produced results. Encouraged, I thrust a hand into the headlamp's open shell and groped feverishly in the tangle of plastic and brass. The fault was located and the offending loose connections bypassed and bound tight with a length of stiff wire and insulating tape. With the engine ticking over and the headlamp shining, the world was a friendly place again, even though my hands were now soaked.

Hour after hour the rain pelted down. At first the only wet parts were my hands, but as I neared Newcastle water began to trickle down the back of my neck again and up into my boots, forced there as I hit deep puddles. My plan was to leave the A1 and travel through the border country on the A68, one of the most exciting main roads in Great Britain. Over the Cheviot Hills it plunges between hills and valleys with such abrupt improbability that while returning from holiday a couple of years ago Eileen, my wife, felt for the first time ever travel sick sitting in the Steib. The Shadow's front wheel was lifting on every peak and the suspension collapsing in every dip with a dizzying swoop. I had hoped to regain a little of this exhilarating sensation on my present journey but I found the relentless rain dispiriting, bewildering even; it just kept on and on, until it was too much trouble even to change my route, so I stayed on A1 all the way to Edinburgh. It was probably a good thing for in that part of the country petrol stations are few, even on A1, and the Ducati's tank held only two and a half gallons. After every 130 miles the search for the next pump would begin. Near Berwick-upon-Tweed, with an almost empty tank and an empty tum, a café, petrol station, and motorcycle shop rolled into one suddenly appeared. The owner of the shop spread my wet things out to dry in front of a hot stove and filled the petrol tank while I had a coffee and snack in the café and attempted to warm up. It was so good coming across this little oasis in the middle of nowhere that I left after 20 minutes feeling almost sorry that I had not been in dire need of a few spares as well.

Welcome lights

The rain was even harder as I rounded the coast to Edinburgh; the city lights were very welcome—for a few miles at least I would not have to ride half blind through the rain. It was about half past six in the morning, just about the time I had hoped to arrive. The two hours lost in repairing the faulty electrical system had been more than made up by the empty and fast A1, which I was very glad to be leaving after such a ride. The Forth Bridge by night is an impressive sight, especially in the rain; reflections and lights give the scene a surrealist outlook, but my artistic reverie was soon shattered by the lone bridge attendant, who when I passed some light, friendly remark as to his likeness, metaphorically speaking, to Horatio on his bridge, treated me to an example of the diplomacy used by sour Scotsmen when dealing with particularly stupid Englishmen. Cheered, I went on my way and passed along what must be the shortest motorway in the world. It

I doubt whether any machine could have covered the distance faster

leads away from the bridge for no more than half a mile before spewing its speeding traffic on to the Perth road. At Perth it was still dark and still raining, so I filled the tank, knowing the road to Inverness was a lonely one, and pressed on, hoping to find a place for breakfast. Just beyond Dunkeld, where the road begins to rise into the Grampian Mountains, the rain stopped and turned into mist, and dawn broke. It was such a relief not to have to work hard at riding that the urge to sleep, despite the cold, became strong and I stopped and turned off the road to run along the river bank in an attempt to warm and wake up, and have a look at a loch or sheet of water I saw not far away, and then I saw a hut. It was dry and warm inside, with some hay and sacks around, so I brought the Ducati along and went inside with my dry change of clothes. The next few hours were sheer bliss. I changed into dry clothes, ate two bars of chocolate, and just sat and enjoyed being out of the weather. Only for a minute, I thought, and lay down in the hay—and awoke nearly three hours later! But I was warm and fit now and within a few moments was on my way.

Beautiful snow

Only a mile along the road snow appeared on the higher ground on either side of the road, and as the road climbed the snow came nearer until the two met. It was beautiful; the sun was shining through in patches, and the snow was clean and new. The road twisted and turned through the Grampians, climbing higher all the while, and the Ducati was going so well that speeds up to 40 m.p.h. were quite safe in the new snow. By now it was about three inches deep and in a few places drifting to more than a foot across the road. My confidence increased too far and 45 m.p.h., even 50, was nothing, until I met a huge lorry coming in the opposite direction almost filling the road and sending a great plume of snow shooting from its wheels. My brakes were useless in the snow and the lorry did not slow one jot. The freezing shock of a faceful of snow at nearly 50 m.p.h. had only one possible conclusion; I fell off. Neither the machine nor myself were damaged, we simply slid gracefully along in the snow, the Ducati on its side. It was all very dignified. Under way once more, albeit more cautiously, I came across the very recent remains of a far more spectacular affair. A Saab lay drunkenly across the ditch; its skid marks followed a long diagonal line across the road, their twists and turns telling quite a story. Not long after that I stopped at a small café for breakfast. It was the sort of thing you read about in books but never actually happens . . . only this time it did. Surrounded by mountains and lying in a wide flat valley echoing to the cries of wild birds, the little café was in a position that any hotel or restaurant owner would consider it worth doubling the bill for, and yet my breakfast cost me only 4s. 6d. A great plateful of bacon topped by a couple of eggs, and with as much coffee and toast as I could eat followed by a hamburger to fill a last little empty corner. I was loath to leave the place and could have quite cheerfully spent another day or two there riding in the snow then coming back for for eggs and bacon!

Soon after I left the cafe the snow thinned and finally gave way to the surface I dread, black ice. What's more, it lay on the downhill stretch leading out of the mountains to Kingussie. I did not actually fall, but that was more by luck than judgment. Gradually the bleak grandeur of the Grampians and Cairngorms gave way to the softer rolling, pine covered hills of the east coast as Inverness approached. I passed through without a stop, wanting the empty road ahead. The east coast of Scotland, I had been told, is uninteresting, compared to the west—but it is not, it is beautiful, and the road to Wick superb, bewildering in its never-ending twists as it climbs and swerves through small villages and around wooded fiords. To the west the great snow-covered towering ridge of the Western Highlands basked invitingly in sunshine. The temptation to spend a day in them, riding along their lonely roads under a high blue sky, was terrific, I promised myself a trip this winter, perhaps on an HT5 trials outfit or something very similar; even the old MSS, come to that, or anything, to explore the hills.

Ducati in its element

From Inverness to Wick the Ducati was in its element. Top speed was immaterial, road holding and braking were of paramount importance, and I seriously doubt whether any machine could have covered the distance much faster, for though the road was damp we flew along. My confidence increased on the wet road until dry-road technique was adopted—as far as I can remember, the only time I have ever been able to do this. Two factors were important here. First, the excellent Pirelli tyres that for the whole of the journey put up a performance equalling our own manufacturers' articles; another "first time" experience for me. The front one was ribbed and the rear nicely rounded, with a good, deep tread that gave me a feeling of confidence merely to look at it. Second factor: the well surfaced, oil and rubber-free road. The low-speed power of the 350 c.c. o.h.c. single proved a great help on the twisting, climbing roads of Scotland; it meant that gear changing, which although initially a pleasure could have become boring after a few thousand of them, was kept to minimum. Bottom gear was very low too low for any but exceptionally awkward bends and pobbling along through the snow. Mind you, in heavy traffic, o the few occasions it was used for suc mundane work, it proved a great ad vantage. Second, third and fourth wer well placed and close enough to cover a conditions, and top made an excellen cruising gear, but it was not so high as t be an overdrive: the low-speed power o the Ducati enabled it to be brought int use at 40 m.p.h. quite comfortably, bu not unnaturally there was not muc acceleration in this gear at low speed I remember my main impression of th little 160 Ducati was to leave well alone but the Sebring is so obviously a fas motorcycle accepting its hobble with a the well-mannered grace of a thorough bred, that I began to wish for more per formance. The hobble in the Sebring case is the tiny carburettor, which althoug contributing towards good fuel econom and low-speed tractibility, strangled an tendency to high revving. I have n doubt, though, that any Ducati deale would be only too glad to supply a mode with the racing Mk. III Ducati's car burettor, a far larger instrument. For a this, it is still a motorcycle capable o breaking the speed limit by at least 1 m.p.h. and, what's more, cruising at tha speed (abroad of course) indefinitely an economically. The only engine char acteristic that could possibly be faulte was a shortage of "flywheel"; couple with the high compression, this meant tha if the revs were allowed to drop as far a the little carburettor encouraged the engine would stall unless free of load After some battery trouble, I found tha the only way to start was either to rol down hill with the exhaust valve lifte engaged, and once the engine backfired drop the lifter, or sit on and be pushe following the same procedure. An attempt at pushing alongside brough about an awkward series of popping and stumbling and lever grabbing. The newe machines, however, have a true flywhee as opposed to the present bob weights.

All-important factor

On a long journey I find that distance shrink into insignificance and time become the all important factor. Leaving Inverness, I realized that Wick was "only" 130 miles away. I translated this into an estimated time/speed of 50 m.p.h. 2 hours 20 minutes. That was it, I wa there, finished. But it is a deflating experience to keep up a mood of this typ for $2\frac{1}{2}$ hours, especially through rain for rain had started again not far beyond Beauly.

Through Wick, and it was only 1 miles to John O' Groats. By 3 o'clock was there, buying a card and requesting the postmistress to sign it (in case th editor thought I had been sunning for a week in Devon). I don't care what the guide books say—the last 20 miles o Scotland are flat and monotonous, and th countryside is derelict, although parts of i are strikingly similar to Land's End, over 1,000 miles south. A quick turn-round and I aimed back down the road to Helmsdale, a beautiful little fishing village by the mouth of the Straith of Kildonan, a fast-flowing salmon river where an hotel had caught my attention on the way up. Fifty miles and an hour later I pulled up in the empty forecourt of the hotel. After a hot bath I found the manageress apologizing because they did not serve dinner—it was out of season—only high tea. My heart sank into my empty stomach, and it must have shown, for the good lady promised high tea would satisfy me. The Scots evidently have the most delightful ideas about tea. It was a huge meal, of soup, and chops, with kidneys, and hot pancakes, and oatcakes. At the end I managed to rise slowly from my chair, then collapsed into another one by a roaring fire.

Sunny and warm

On Friday morning I left at 10 o'clock, hoping to reach London in 12 hours. It was sunny and warm and I looked forward to a good dry run home, and this time I would take the A68 over the border country. But it was not to be. Instead of taking the lower road I had used in coming up, I decided on the more interesting A939 from Inverness to Perth. It runs right over the top of the Cairngorms and Grampians through Tomintoul, Braemar and the Devil's Elbow. It is a wild, high and lonely road, and I paid for the pleasure of using it. Twenty miles beyond Tomintoul I met a chap with a puncture in his car tyre; he had no spare. Sitting side saddle with the tyre on his knees he came with me on the Ducati back to Tomintoul, where after an hour a local garage repaired it. Back we went, but he had no jack. A great rock had been forced under the axle and the wheel worried from its bolts; with the tyre inflated and no pump the brake drum was three inches lower than the wheel. One of us bounced the car side ways on to a larger rock, while the other jammed it further into place. I left him thankfully, but it was 2 o'clock now and my time schedule was ruined. But it was still sunny and the scenery was magnificent so I did not worry unduly.

High, desolate spot

Everything went well until I reached the Devil's Elbow, a high desolate spot even on a sunny day. The bike started to misfire and slow; I suspected a blocked jet but it was nothing to do with the carburettor. However the bike started so I rode off, and just as it was petering out again my " puncture " friend came past in his Humber. I waved frantically. He waved too and motored on. The battery appeared to be flat, but the Ducati would not start even from the alternator after a long downhill run, so I felt stumped. Luckily a Land Rover came along and gave me a lift into Blairgowrie, about 40 miles along the road. I bought an Ever Ready 6 volt lantern battery and thumbed another lift back to the bike, a journey I

DUCATI SPECIFICATION

Engine
  Capacity, 340 c.c.
  Bore and stroke, 76 × 75mm.
  Compression, 8.5 to 1.
  Alloy head and barrel.
  Roller bearings throughout.
  Two valves operated by o.h.c., driven by bevel gears and shaft.
  Timing, inlet open 20 degrees, close 70 degrees. Exhaust open 50 degrees, close 30 degrees.
  Carburettor, Dellorto 24mm.
  Filter, dry felt.
  Silencer, steel baffles.

Transmission
  Primary transmission by helical gears.
  Clutch, 13 plate running in oil, cable operated.
  Gears, five, operated by offside rocking pedal.
  Sprocket: engine, 26; gearbox, 18; clutch, 57; rear wheel, 45.
  Final transmission by $\frac{1}{2}$ × $\frac{5}{16}$in chain. Adjustment by sliding wheel spindle.
  Shock absorber in rear wheel.

Frame
  All welded. Single tubing throughout, open at base with engine support.

Suspension
  Rear. Bronze bushes supported in engine plates, with steel spindle. Three-position load adjustment in spring and two-way damping units.
  Front. Telescopic fork. Soft one-way damping.

Brakes
  Front, all-alloy 180mm s.l.s., cable operated.
  Rear, all-alloy 160mm s.l.s., cable operated from nearside foot.

Wheels and Tyres
  Both Q.D. 18in steel rims, with 18 × 2.75in ribbed front tyre and 18 × 3.00in studded rear by Pirelli.

Fuel Tank
  Steel construction. Bolted to frame. Capacity 2$\frac{1}{2}$ gallons.

Instruments
  Viglia speedometer only, driven from front wheel, illuminated, undamped needle.

Electrical System
  Ducati crankshaft-mounted alternator and control box; 60W output.
  Ignition, coil and contact breaker.
  Lighting: 6in headlamp, pre-focus 35W × 30W bulb. 6W tail × 18W stop lamp.
  Centre and prop stands.

Dimensions
  Wheelbase, 53in; seat height, 35in; ground clearance at prop stand, 4in; at crankcase 5$\frac{1}{2}$in; overall length 78in. Weight, 271lb.

did not enjoy for Blairgowrie was warm and comforting, the Devil's Elbow cold and bleak. A mist had fallen and night was only an hour away. With the dry battery fitted the engine started first kick and we roared off to Blairgowrie where the old battery was tested and pronounced o.k. But as soon as it was fitted it went flat. After much sleuthing the garage found an internal short that only showed itself when the engine was started and vibrated the battery. A phone call to a dealer in Perth, who kindly agreed to wait open for me, and found a good secondhand battery. Feeling confident once more, and very grateful to the dealer for his trouble, I pushed on towards home.

From Perth nothing untoward occurred. It rained, of course, but that was normal by now and scarcely worthy of a second thought. The Ducati's exhaust droned and the rain fell steadily. Until now nothing had passed the Ducati other than a Police car and a Mini Cooper S scuttling along like a bandy crab on its extended wheels, but then with muted roar a Triumph Bonneville went rocketing past. The rider, Barbour-suited and luggage laden, gave me a cheery thumbs up, and with that gentle undulating movement common to all fast-moving vehicles disappeared into the night and left me feeling all the more lonely. Did I know him I wondered? The world of long-distance riders is a small one, no more than a few hundred I would wager, and this man was travelling air on his Triumph, I could tell. His style was set and steady, urgent and yet relaxed.

Stopping at all the cafés along the A1 to lengthen the journey, for I did not wish to arrive home in the early hours of the morning, failed to break the gentle yet persistent tug to sleep, and so somewhere on the road I found a comfortable looking bridge, pulled the bike behind a pillar, and slept. I awoke to the sound of a dog snuffling in my ear and a voice apologizing for waking me but was I all right? A policeman had seen the bike and body and thought he had better investigate. I arrived home at 7 a.m. demanding an enormous breakfast and afterwards, feeling pretty chirpy, went to bed; awoke at lunch time (sheer instinct) and spent the rest of the day lounging comfortably with a good book and listening to the rain outside.

A motorcycle with far fewer faults than most and capable with very little modification of entering most spheres of competition, whether racing, trials or scrambles. A worthy successor to the Gold Star. For a man who wants his machine to do everything reliably and with the minimum of fuss and bother.

JOTTINGS

Distance covered: 2,089 miles.
Land's End to John O'Groats: 1,025 miles.
33 gallons of fuel used = 63.10 miles per gallon.
Two pints oil used.
Average speed: 35.10 m.p.h.
Average speed excluding sleep and repairs: 51.5 m.p.h.

PERFORMANCE

Speedometer accuracy: 4 m.p.h. fast up to 60 m.p.h. Beyond this speed needle movement prevented accurate speed evaluation.
Time taken for quarter-mile standing start: 17.40 seconds (best of dozen runs).
Speeds in intermediate gears: unable to evaluate owing to exaggerated needle movement.
Approximate top speed in fourth gear: 80 m.p.h.; in fifth (top) gear, with rider in light clothing and crouching: 85 m.p.h. (in heavy riding clothes, 80 m.p.h.)

MM TRACK TEST

DASHING DUCATI

CHARLES DEANE TESTS THE 350 MINTER AND EVERETT WILL RIDE IN THE 500-MILER...

The overhead camshaft 350 cc motor of the Ducati Sebring Mark 3 pokes out 29 bhp at 8,500 rpm. In unit with a five-speed gearbox, the motor is set in the lightweight frame used on the 250 Daytona

The front stopper on the Sebring is a 180 mm. single-leading shoe unit and although perfectly adequate for road use, was found to suffer a little under high-speed racing conditions around Brands

The rear suspension was a little on the soft side for production racing, but perfect for road use. However, this will be corrected in time for the 500-miler. The rear brake unit is cable operated

With chest and chin pad strapped on for "getting down to it", the Sebring Mark 3 has a standard 3½ gallon fuel tank fitted. Oil is contained in the sump of the motor. Note the quick-fill petrol cap

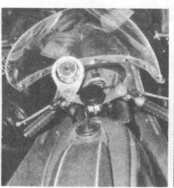
The handlebar layout is neat and uncluttered. Brake and clutch are adjustable at the handlebar levers and the ignition switch is set in the headlamp shell. Rev counter should be marked in red at 8,500

I wonder where that 50 went? The 29 mm. Dellorto carb with large air intake had the choke slide removed for Brands. It kept closing at high revs and caused misfiring till removed

If you're going to watch the Brands Hatch 500-mile Production Race, keep an eye on the leader board in the 500 class for Derek Minter and Reg Everett on the Vic Camp entered 350 Ducati!

Having recently ridden the Ducati at Brands during the Wednesday practice sessions and also having watched Charles Mortimer Jnr. romp the roadster around the Kent circuit at consistent 1 min. ? sec. laps, I imagine a few of the riders of the bigger bikes will be in for a surprise. Charles Mortimer's comments were that the rear end was a bit soft and that the front brake could do with more bite, but these are both things that will be right on the day.

Keeping the engine buzzing at around the 8,000 rpm mark and making use of the five-speed box, I found that the Ducati was a match for all but a few of the fastest racing 250s. The handling couldn't be faulted, except as Charles said, a slightly soft rear end caused very slight pitching on the bumpier bends.

Top speed

Top speed of the machine was not really obtainable on the short Brands Hatch straights, but with well over 90 being indicated by the fluctuating speedo needle and with the knowledge that Minter has tested the bike at the circuit with a 1 min. 3 sec. lap, an approximate 105 to 110 mph would seem about right.

Starting the bike was perhaps one of the most difficult tasks and although the motor had only a 9.0:1 compression ratio, I could literally jump up and down on the kickstart three or four times before it would budge. However, with the motor running, there was virtually no vibration and only the rev counter told you it was time to swap the cogs.

With the rearsets set high on the frame, I was unable to ground them, but there were signs where Minter had been riding and I understand he wants them raised another inch. There's no doubt that the Dunlop triangulars provide a fantastic angle of lean in safety.

The big question with any roadster entered for production racing is how long will it last?

In the case of the Vic Camp Ducati there should be no problem, because the bike had already covered over a thousand high-speed test miles without a failure and the engine, apart from a few oil stains, was clean and untouched.

The only modification which Vic Camp had made to the Ducati was to fit a megaphone-type silencer, which although not as raucous as the racing megaphone, still gave a healthy bark under full throttle. However, he is not too worried if he has to revert to the standard unit as there was no great improvement in the performance with the baffled megaphone unit.

Forecast doom

Electrics on the Sebring Mark 3 are standard with coil ignition and as we were only able to ride the machine on the track during daylight hours, we can make no comment about the efficiency of the lighting system.

Many have forecast the doom of the single-cylinder medium and large capacity motorcycles, but Ducati have been proving critics wrong in recent Italian race meetings where Spaggiari has been dicing with and on occasions beating the multis on his single-cylinder, desmodromic Ducati.

We understand that the company is now going to market roadster versions of the desmo racer, using the mechanical valve gear which opens and shuts the valves without springs.

Red line raised

The cylinder head is the only unit which requires changing and it is a straight swap from standard to desmo valve gear cylinder head. This immediately raises the red line on the rev counter from 8,500 to over 10,000 rpm.

The standard 29 mm. carburettor is also replaced by a larger bore unit of 32 mm. to provide improved breathing at the higher revs.

There's no doubt that if these machines are marketed in this country, they will provide a much-needed boost to the middle capacity classes and will give some of the bigger banger enthusiasts a few shocks on the performance side.

Vic Camp's aim is to win the 500 class in the 500-miler and with "The Mint" on board the Ducati, I reckon he stands a pretty fair chance of doing just that! ●

Ducati expert Bert Furness dismantles the front end of this race-proved lightweight

The first step is to unscrew the top caps with the correct size Allen key. If they are on the stiff side, they might need a tap with a mallet to free them

By waggling the clip-on bars before undoing their bolts, the fork legs can be off in readiness for the part of the strip op

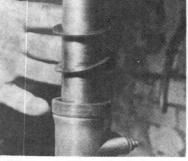
When rebuilding the forks, do not omit the two washers on each leg. Generally, it is best to replace the lower washer, as it is made of fibre and rather thin

When removing this Allen the damper securing block tend to move. If it does, length of rod, filed to s driver shape, to hold the

▶ **Overhaul a pair of Ducati front forks and you'll probably spend no more than a few shillings on a couple of oil seals and fibre washers!**

There are no bushes to worry about. In effect, the bottom fork cover is simply a huge fork bush made to such fine tolerances that, given proper lubrication, it should last the life of the machine.

Obviously with such a large working area, wear and sloppy fork action is down to a minimum and steering troubles on these high performance four-strokes are very rare.

As with most other forks, first step should be to undo drain plugs and eject all the by applying the front bra depressing the forks to the li of their travel a couple of tir until the flow stops.

All spanner sizes are met including Allen keys, so do try to "make do" with Brit spanners—you'll only butch the nuts.

With a large Allen key, top nuts can usually be u screwed without much troub On assembly, be sure to get t thread started squarely bef you tighten.

DUCATI

 he pinch bolt can be slack- off to allow fork leg to ed out. For this, the special tool is very useful, but is not available . . .

 . . . the spindle can be replaced in its bottom locating hole and tapped carefully with a soft-ended mallet. By tapping gently on either side, the leg will free

 Stand the leg upright in the vice and lift off spring. This will show no wear. In fact the only wear on these sturdy forks is usually confined to oil seals

 A "C" spanner is needed to re- move the dust cover. Run some penetrating oil down the threads if removal proves difficult. Once started, unscrew by hand

 damper rod and spring can be lifted out and inspected. a tumble involving serious damage all parts should be ughly checked by a dealer

 Never disturb the damper as- sembly, which is retained by a circlip in the fork leg. This component can be difficult to refit, and should last life of forks

 Prise out the oil seal with a screwdriver. When fitting the new seal, it is best to oil it slightly and tap it in with the palm of the hand to avoid damage

 Before rebuilding the forks, make sure that all parts are cleaned and threads on alloy legs are not damaged. Each leg needs approx. 100 cc of SAE 20 oil

If the caps are left on the garage floor they can very easily lose the lead thread, therefore if they are likely to be off for any length of time, protect those threads.

The clip-on handlebars are ready-made wrenches when it comes to freeing-off the fork legs. By waggling them to and fro the most stubborn legs can be freed.

Not many private owners will have the special tool for tapping out the fork legs, but this is easily overcome by replacing the front wheel spindle and tapping it gently with a soft mallet.

An ordinary "C" spanner re- moves the dust caps but again, in the absence of the proper tool, try wrapping a piece of fine emery around the cap and grip- ping it with your bare hands. You'll be surprised at the amount of grip the emery gives on the smooth paintwork.

Removal of the cover will expose two washers. The top is of metal and the lower of fibre. Renew the fibre one as a matter of course.

Now we come to the small Allen bolt in the damper securing block. Sometimes this will come out with no trouble, but if the block begins to turn, you will have to get a long piece of rod about $\frac{1}{4}$ in. diameter and file it to locate in the notch in the block itself.

This should hold it still until the screw slackens off.

Visual tests will soon tell of any damage to the fork legs or any component parts. By using a straight-edge or a piece of string held taut alongside the suspect member all but the most minute distortion will become apparent.

The damper unit will require no attention and should never be disturbed. Under ordinary conditions this unit is subjected to the very minimum of wear and will last the life of the forks.

Oil seals and all Ducati spares are available from the Ducati specialists, Vic Camp Motor- cycles, 131 Queens Road, Walthamstow, London E17.

If you prefer it, they will do the job for you, but there will probably be a waiting period in the summer months when all repair shops get busy.

Lastly, when fitting oil seals they should be lightly oiled and tapped in gently with the palm of the hand. They are precision parts and damage easily.

FORK STRIP

By removing three Allen screws access is gained to the overhead camshaft which is driven by the helical bevel gears shown

The sump is finned for better cooling. Checking oil level is easy—there is an oil dipstick on the end of filler plug. Sump holds 4½ pts.

Contact breaker points are kept dry by an 'O' ring behind the cover plate. As you can see, they are very accessible and simple to set

A permanently attached h with which to adjust the rear pension to suit varying loads away with the need to carry a

DUCATI 250s

When you read the e specification of a 2 Ducati it seems more lik of a quality racing mac This is hardly surprising a Ducati factory has bui name on the successes ohc lightweight singles.

The fact that the moto barely changed over the is proof enough that the m and owners of this thoroug marque are well satisfied both performance and dura

We picked up two bran machines from Bill Hannah of Liverpool. The standard "Monza" and the road-g racer called the Mark 3.

Both machines had to b in—a tiresome task on any performance machine, espe a machine like the Mark 3 v seems to want to gallop at opportunity. Nevertheless set to it and the miles mounted up until we were to take the safety stop out o Monza carburettor and give machines their head.

STARTING PROBLEM

As I have said, the mach were new and had not ridden, so at first starting a hit and miss affair. In fact had been asked to commer the ability to start either mac on the first day we rode th would have been put insid using abusive language worse!

However, after a few d when the engines had been for several hundred miles, s ing improved. The "Mo

SPORTSTE

you can see the large diam- spokes and high quality cast- on alloy front forks and 7 in. Note vents front and rear

On the Monza a sturdy prop-stand is fitted. The special attachment allows it to be used without effort while sitting astride the bike

Handlebar layout of the Mark 3 is similar to the Monza except for the tachometer and rounded head- lamp shell. Both have a damper

Unmistakably the Mark 3. Open bellmouth of the TT-type carb and rev-counter attachment are the giveaway signs, all-else is the same

starting technique was simply to turn on the petrol and kick the engine over without prior flooding. This usually meant a second or third or fourth kick start.

I found that flooding was necessary on the Mark 3 and as there can be no tickover until the motor is really hot the throttle had to be opened to give about 2000 rpm immediately. This was due to the TT type Dell'Orto carburettor on which there is no throttle stop screw.

The rev-counter starts at 2000 rpm and some discipline was needed to slip the clutch and keep the rpm up when approaching road junctions. If the revs dropped, the motor died and often this meant a "bump start" in a busy high street.

HANDLING

The American-style handlebars were not to my liking at speeds above 60 mph or so, but they were comfortable enough at low speeds in town and the general handling was superb, as one would expect from a race-bred machine. Whatever the road surface or weather conditions the well balanced suspension inspired confidence and the angle of lean possible was adequate for all normal riding.

Performance figures for the "Monza" were somewhat dis- appointing—a genuine 73 mph (electronically timed) was the best we could attain although the speedo was showing an optimistic 87 mph!

Petrol consumption was another story—throughout the test the "Monza" returned an average of over 75 mpg and the Mark 3 was not far behind, averaging about 70 mpg. A truly remarkable figure for a machine with a 100 mph potential.

Throughout the test, both engines remained completely oiltight and no oil was used at all. One important thing to remember was that the petrol tap should always be turned off when parking the machine. If you omitted to do this, petrol would run down the carburettor (downdraught type) and into the bore, thus making the plug wet and starting impossible.

CLUTCH, GEARS, ETC

The rocking gear pedal gave a slick positive action and was light and easy to use—providing you didn't try to use it like an ordinary gear lever.

If you tried the lazy method— and I did—the rubberless lever soon damaged the top of shoes, etc, when hooking the lever up with your toes while changing down.

My only criticism of the heel and toe method was that the footrest was positioned too low and the foot had to be lifted off its rest to change gear. Apart from that, cog-swapping was simple and positive at all times.

The clutch needed adjustment several times on both models during the first few days until it "settled in". Once the initial running in period was over, the unit seemed capable of standing up to anything we could hand out—sprint starts and vicious engine braking with no signs of slip.

Fifth gear, like most fifth gears, was something like an overdrive and was never used about town, although it no doubt contributed to the low petrol consumption figures on long, fast runs.

Routine adjustments were easy to carry out as everything was easy to get at. Two screws hold the points cover and adjusting only took about two minutes. A dip-stick on the sump filler plug enabled oil to be checked at a glance. Clutch and brake cables had finger adjusters but neither brake needed attention while the bikes were in our possession.

Braking was very good, with both machines being brought to a standstill from 30 mph in 30 ft., by the 7 in. front and 6½ in. rear units used together. Even from 90 plus mph there was no sign of fade.

SPARES SERVICE

Often the snag with buying a foreign machine is availability of spare parts. This is not so with the Ducati. Bill Hannah can offer a 24-hour spares service and has even sent one of his men to the USA to get really "with it" on the spares side and consequently is able to give a first-class service with no annoy- ing hold-ups.

Lighting was all right for about 60 mph after dark, but I think that for a machine with the performance of the Mark 3 you really need a 12-volt system. The sealed beam lighting used on the Monza gives a greater intensity of light than the con- ventional detachable bulb type but, where you have two fila- ments and therefore two chances if one filament blows with an ordinary bulb, should you crack or damage the sealed beam type of light—you have no head- light at all! The light is, in effect, simply an over-sized bulb!

For touring with this sort of equipment it would pay to fit some sort of stone guard—just in case.

No horn is fitted on the Mark 3, so to comply with the law a bulb horn must be used. The Monza, however, is fully equipped with a battery and coil ignition, but no battery or ignition key is needed on the Mark 3.

Overall finish was extremely good. The only signs of rust appeared on the Allen screws at the base of the camshaft tunnel and on the silencer holding bracket. Paintwork was excellent as were the alloy brake castings.

Passenger comfort was reason- able on the Monza but the narrow dual seat, coupled with firm suspension on the Mark 3, did not allow much pillion comfort. After 50 miles or so, the res- tricted movement imposed by carrying a pillion passenger induced saddle soreness on the rider too. Whereas far longer journeys were covered alone with no discomfort.

Performance was in the big bike class and the motor never missed a beat, the faster the

ROADSTER

MM compares these two Italian lightweights from a racing stable

DUCATI 250s... MONZA v MARK 3

DUCATI 250 SPECIFICATION

MONZA

Engine:
249 cc (bore and stroke—74 x 57.8 mm.) overhead camshaft single cylinder. Roller bearing big-end with crankshaft supported in two ball bearings. Plain small-end. Compression ratio 8 to 1. Lubrication by dry sump. Oil capacity 4½ pints.
Carburation:
Single Dell'Orto, 24 mm. diameter choke. Air slide is operated by control lever on right handlebar. Air filter fitted.
Transmission:
Helical gear primary drive; secondary by ½ × 5/16 in. chain. Multi-plate clutch with steel and composition plates running in oil. Five speed, foot operated gearbox giving ratios 15.8, 10.8, 8.4, 6.87, 6.06 to 1.
Electrical Equipment:
Ignition by coil. Charging by 60 watt alternator through rectifier to 6 volt 13.5 amp-hour battery. Headlamp 6 in. sealed beam unit.
Suspension:
Telescopic front fork, hydraulic damping. Swinging arm rear with three-position adjusters.
Brakes:
7 in. diameter front, 6½ in. rear.
Capacities:
Oil—4½ pints.
Fuel—2¾ gals.
Dimensions:
Wheelbase, 51 in. Saddle height, 31 in. Ground clearance, 6 in. Weight, 275 lb.
Price: £275 17s. 0d.
Agent: Bill Hannah Ltd, Liverpool.

MARK 3

Engine:
249 cc (bore and stroke—74 x 57.8 mm.) overhead camshaft single cylinder. Roller bearing big-end with crankshaft supported in two ball bearings. Plain small-end. Compression ratio 10 to 1. Lubrication by dry sump. Oil capacity 4½ pints.
Carburation:
Single Dell'Orto, 29 mm. diameter choke. Air slide operated by lever on right handlebar.
Transmission:
Same as Monza above.
Capacities:
Oil—4½ pints.
Fuel—3½ gals.
Suspension:
Same as Monza.
Electrical Equipment:
Ignition by alternated current ignition coil supplied by 6 volt 40 watt flywheel alternator. No battery fitted.
Brakes:
7 in. front, 6½ in. rear.
Dimensions:
Wheelbase, 53 in. Saddle height, 31 in. Ground clearance, 6 in. Weight, 247 lb.

Speeds in gears; MONZA:

	Minimum	Maximum
1st	6	28
2nd	12	40
3rd	16	52
4th	20	63
Top	25	73

Speeds in gears; MARK 3:

	Minimum	Maximum
1st	8	33
2nd	13	52
3rd	18	70
4th	23	85
Top	30	92

Price: £281 7s. 9d. inc. P.T.

Value for Money

going, the better it liked it. The highest speed reached by anyone on the staff was 92 mph, but by fitting the racing megga, clip-on bars and jetting up, the maker's claimed 100–110 mph does not seem unreasonable.

While riding a Ducati you get the impression that a lot of thought has gone into the design. For instance, the extension to the side stand on the Monza enables the machine to be parked without the indignity of a balancing act while you try to pull the hidden stand from beneath a dirty motor. Then there is the sensibly-sized toolbox and the extensive use of Allen screws instead of those infuriating cross heads.

The spokes on both wheels are large diameter, also the control cables (inner) are about the thickest made. At Bill Hannah's they say they have never known one to shed its nipple.

The Mark 3 at £281 7s. 9d. is very good value for money. It is undoubtedly a machine for the high performance lightweight enthusiast. The Monza, at £275 17s. 0d. is relatively docile and more suitable for commuting, although by no means out of place on the motorway.

Once we had mastered the art of starting the beasts, we at MM came to like and admire these cammy 250s and could find no serious faults with them. Possibly the lighting could be improved and the gear lever or footrests reset to make gear-changes easier. Apart from these small things, the Ducati is a true Italian thoroughbred lightweight.

Recent successes in the Thruxton 500 Mile Race and the 250 cc Production TT bear witness to the speed and, what is perhaps even more important, reliability of this ohc sportster. It's no use blowing them all off for a couple of laps and then disintegrating!

The old saying "Racing improves the breed" was never more true than in the case of the Lightweights from Ducati Meccanica.

PHOTOGRAPHY: BENNO FRIEDMAN

CYCLE ROAD TEST

DUCATI 350 SSS

SSS stands for "street-scrambler-sport", and the big Ducati is every bit as versatile as its name would have you believe—and much more sophisticated.

It's not the kind of a motorcycle that gets a lot of attention, the Ducati. Pull into your local Macdonald's and wing her a few times, and, admit it, not that many heads turn. It doesn't have the kind of dressed-to-the-teeth chrome and enamel ostentation of, say, a Sportster, nor the screaming all-new-for-'69-ness of a BSA-3, nor the authoritative snarl of Triumph Bonneville. Ride down the street on your Ducati 350 SSS and chances are pretty good that the local queens will leave your clothes alone.

Unless, that is, the local queens know a lot more about motorcycles than they're usually given credit for.

The Ducati 350, see, is very much an afficianado's motorcycle, replete with such subtleties as are often wasted upon novitiates and upon those who appreciate topically and from afar. A Velocette, it is, if you can picture an honest-to-God 1969 Velocette—with manners (which is to say, one that starts). It's tough, smooth, beautifully designed, plenty fast (90 mph), and more modern, probably, than you could imagine a single-cylinder motorcycle to be.

It's also legitimately dual-purpose. You may recall a few years ago when Ducati was in the throes of an identity crisis: they started out with a 175, evolved to a 200 and then up to a 250. The following year it was sleeved and destroked and offered as a 160 (as well as, of course, a 250). Then, in 1967, its displacement was increased to 350, and, lo and behold, the Ducati Sebring was upon us. Followed by the 350 Mark3, essentially the same as the Sebring but with enough performance-oriented changes to make it a good 25 mph faster than its predecessor.

And then the 350 Motocross in 1968, for the most part identical to this year's SSS. All those Ss, by the way, stand for "Street-Scrambler-Sport," which should pretty much cover all bases. Except T, for trail, where the 350 is as comfortable as it is on the street. More on that later.

The fundamental engine design utilized by Ducati for all these years, single-cylinder shaft-driven overhead cam, is carried over to the SSS—and it's as sophisticated a design as there is on the market. Barely oversquare (76 mm bore, 75 mm stroke), high-compression, primary drive by helical-cut gears, a five-speed transmission. The overhead cam, though, is the snapper. Ducati does it differently than, say, Honda, by using an externally-routed vertical shaft to drive the cam instead of a chain. The shaft has a lot going for it: there is no need for a tensioning device, cam timing isn't likely to vary with age (chains tend to stretch), and, because it is driven by a helical gear on the timing shaft and drives another helical gear on the camshaft, it is perfectly silent—no need for rubber-damped drive sprockets. And the Ducati's cam-drive shaft is more than adequately sup-

DUCATI 350

ported by three large-diameter ball-bearing sets: one just above the lower drive gear, one below the cam drive gear, and one in the middle.

Another benefit of a shaft-driven overhead cam is ease of maintenance. Ever tussled with the upper-end of a chain-driven overhead cam engine? Especially one which employs an endless chain (no master link)? If the manufacturers of such are trying to discourage the average owner-tinkerer from tinkering, they succeed beyond their wildest expectations. What with losing the timing chain down in the crankcases, and having to rivet the chain back together after re-installing the cylinder head, and fussing with the chain tensioner, your run-of-the-mill bike nut would just as soon let somebody else do it next time. And what fun is that?

No problem with the Ducati. Although the fuel tank might have to be removed for more elbow-room, everything else is a pushover: intake valve adjustment takes place under one easily-removed cover, exhaust valve checking under another, and cam timing alignment under a third. The camshaft itself is supported in the aluminum cylinder head by two circlip-secured ball-bearings, and operates the valves through two plain-bushed rockers. Valve adjustment on the 350 is handled by locking screws in the valve-actuating ends of the rockers.

The Ducati's valve springs are of the mousetrap variety, two springs per valve—one on either side of the valve stem. The valve springs do not, as you might suspect, utilize normal upper and lower collars. Instead, the lower collar is a stamped-steel affair with openings at either end to accomodate the stationary ends of the springs, while the upper collar is shaped somewhat like a yoke and secured to the valve stem by normal keepers. Pressed-in valve seats are used, as well as rubber oil seals around the valve stems.

And, thanks be, the Ducati is also equipped with a compression release, operated by a lever on the left handlebar. Without it, the SSS is a bit of a bear to get lighted first thing in the morning. A 350 single isn't exactly small, remember, and, with a compression ratio of 10:1, not amenable to half-hearted kick-throughs. But it's ignition system (battery-and-coil) is more than adequate, and if the starting outline is followed, persuading the Ducati to life is seldom more than a two- or three-kick affair. Here is the outline: turn on the fuel. Make sure that the kickstand is down (if it's not, the kickstarter becomes snarled up with that device, causing the kicker to feel like a boob). Tickle the float until fuel begins to seep out. Turn the engine over until the piston comes up on compression, and then, pulling in the compression release lever (which, by the way, lifts the exhaust valve off its seat), ease the piston past Top Dead Center. Let go the compression release and kick away.

(It's difficult to convey the importance of the compression release. In our haste to get underway at one point during the test we neglected to use the release, and discovered that trying to start the Ducati without it is like trying to milk a cow that isn't hip: you'll get kicked every time.)

A five-speed transmission never hurt anybody, and it does wonders for the SSS. Its internal ratios are 2.53-1, 1.73, 1.35, 1.10, and 0.97 in fifth: a gear for every purpose. As with most legitimate street scramblers tested, we subjected the SSS to the drive to the scrambles course, a couple of hard hours of scrambling, and the drive back to the shop. In spite of the fact that the 350 carries a "street" drive sprocket combination we never felt that the bike was over-geared for scrambling; indeed, the Ducati felt adequately geared for all but the slowest and roughest New England-type courses.

Which is understandable, considering the power characteristics of the engine. The big single makes horsepower at any

reasonable rpm, and, although it doesn't generate the kind of steam necessary to head the CZs off at the pass, we found it plenty strong enough—and a leadpipe cinch to ride. The hotter the two-stroke the narrower the powerband, and the more abrupt the transition from no power to maximum power. A handful to ride, right? Should the transition occur in the middle of a turn, down you go. The Ducati 350 is gentler than that, more predictable, and, for all but the Expert, an awful lot easier to live with.

The Street-Scrambler's frame is similar to that used on the 350 Sebring—only longer. The steering head is located by a large-diameter backbone tube and a large-diameter downtube. The backbone tube curves downward and serves as the main rear engine mount; the abbreviated downtube (it doesn't loop completely under the engine) supports the front engine mounts. So the engine, then, acts as a load-carrying part of the chassis. All frame member junctions are heavily gusseted, especially those in the region of the swing-arm pivot point. It's a good, stiff chassis.

And the front forks are good, stiff forks —the only feature of the Ducati that we didn't particularly care for. They'd be perfectly satisfactory on a roadracer, or even a pure street machine. But they're too much for playing around in the dirt, transmitting a lot of front-wheel pounding to the rider's arms and shoulders. The rear suspension units are much better. They're adjustable, offering three different degrees of stiffness. And you won't need a special ring spanner to do the job, either—the shocks are equipped with heavy wire adjusting clips, and each of the three positions are clearly marked. We found the "minimum" setting to be most satisfactory for the dirt.

The rear suspension units are also protected from dust and other abuse by rubber gaiters, similar to those used on the front forks. You see them and you have wonder: why doesn't everybody do it? After all, the rear shocks are at least as susceptible to dirt as the front forks—perhaps more so. The only other motorcycles that we can remember having rear gaiters were the I.S.D.T replica MZs, imported several years ago and costing close to $1500.

The Ducati works well as a street machine; it works very well as a scrambler; but it should fairly sparkle as a trail machine. Fitted up with an oversized rear sprocket, the SSS looks to be nearly unstoppable. The engine cranks out a goodly amount of low-end grunt, it's bashplate is stout enough to be more than ornamental, and with a 3.30-19 tire on the front and a 4.00-18 on the rear, ground clearance is an altitudinous 8¾". And it's a beautifully balanced motorcycle, light on its feet, and equipped with one of the most comfortable seats around—no small consideration, if you happen to be a trail-basher. Nor is fuel tank capacity a small consideration. Many street-scramblers have been introduced within the last few years with ultra-small tanks, paying the price for "style" in the coin of practicality. If any kind of motorcycle needs range, it's the trail machine.

DUCATI 350

Good for Ducati—their SSS carries just under 2½ gallons, which ought to be plenty.

Every now and again a motorcycle comes along that surprises us—either it's a lot better than we thought it would be, or a lot worse. The Ducati, really, was no great surprise. They've been making fine motorcycles for some time now, all well designed, all good performers, and all quietly understated. So take your eyes off all those charisma-cycles for a second and give the Ducati a look. If you really know motorcycles and are sensitive to mechanical finesse, the 350 SSS is bound to turn you on.

DUCATI 350 SSS

Price, suggested retail	$839, FOB N.Y.
Tire, front	3.50 in. x 19 in.
rear	4.00 in. x 18 in.
Brakes, front	7.09 in. x 1.34 in.
rear	6.30 in. x 1.34 in.
Brake swept area	56.2 sq.in.
Specific brake loading	8.5 lb/sq.in.
Engine type	4-stroke SOHC single
Bore and stroke	2.99 in. x 2.95 in., 76mm x 75mm
Piston displacement	20.75 cu. in., 340 cc
Compression ratio	10:1
Carburetion	(1) 29mm, Dellorto
Air filtration	Paper element
Ignition	Battery & coil
Bhp & rpm	N/A @
Fuel capacity	2.38 gal.
Oil capacity	6.4 pt.
Lighting	Alternator 6V, 70 watts
Battery	6V, 13.5 ah
Gear ratios, over all	(1) 16.73 (2) 11.72 (3) 9.14 (4) 7.49 (5) 6.56
Wheelbase	55 in.
Seat height	31½ in.
Ground clearance	8.75 in.
Curb weight	303 lb.
Test weight	478 lb.
Instruments	Speedometer/odometer
Standing start ¼ mile	16.6 seconds—74.10mph
Top speed	90mph (est.)

RAY KNIGHT RACE-TESTS the DESMODROMIC DUCATI

IN THESE days of the developing two-strokes, when every other model on the road seems to be one, you might think that they would also dominate production racing in their classes. Perhaps they should, after all the advances that have been made in the last few years. Sure, I know that your actual Suzukis, Bultacos and Ossas have done a fair amount of cleaning up in the international events, though a Ducati did win the 250 Production T.T. this year, which brings us nicely to a point of this particular piece—a Ducati.

The subject of this racer test might well be described as the four-stroke's answer to all the aforementioned progress—at least in club circles I reckon that it will be for a while at least. But to set the scene; in club production racing, although the two-strokes have been predominant in terms of numbers over the last two or three years, at least in my part of the world it's usually been a Ducati that has won the 250 class. Usually, too, it's been Clive Thompsett on his 1965 Ducati Mk I in pretty standard trim that has kept the Suzies and the rest at bay. So if a better model is produced, then it obviously has to be capable of blowing this one off without too much trouble, or it's hardly worthwhile.

So what is the model that we have to put to the test? It's basically the same, single-cylinder model whose ancestry can be traced right back to the 125 cc Grand Prix racer of the early 50s, that grew to a 175, became a roadster, a 200 and then a 250, and now has grown a "desmo" head. In case you don't know why that should be a good idea, having the valves opened and closed mechanically, without the aid of valve springs, means that the energy used to overcome the pressure exerted by them can be put to better use. It also means that valve bounce at high revs is no problem, you can run at higher revs—get more fuel in and out and produce more bhp=mph. But does it? That is what we were out to prove.

The scene of the action was to be Snetterton and the meeting was Bemsee's Guinness Trophy, September 28. To make sure the results obtained were really conclusive, Clive's Mk I was in the same race so this test would prove whether the added complications of desmo valve gear were really worth-while.

So where do we get our test model from? Well, during the Hutch I spent about half an hour dicing with Charles Mortimer round Brands Hatch. Charles was riding Vic

As quick as many a machine of a larger capacity, the Ducati overhauled more than its fair share of opposition. Here, with Ray Knight's long frame tucked neatly behind the screen, it takes the 650 BSA of J. R. Roberts.

Camp's '69 desmo model and after the race we were discussing the relative performances of the "Duc" and my Daytona, and Vic suggested that I ought to have a ride. I wasn't going to argue with that and another racer test was arranged.

When Vic produced the bike I fully expected to have some difficulty in fitting my 6ft 1in round it, and although it looks no more than a 250, to my surprise I dropped naturally into a racing crouch that got most of me behind the fairing and gave ready access to the controls. It felt like a racer and I knew from past experience that it went like one.

Now the bike had done rather a lot of mileage before I got my hands round it. The T.T. 500-Miler Barcelona 24 hour and various club events. Not to say that it was untouched after all this, but no major rebuilds had been carried out. One thing that had been changed, not unnaturally, was the front brake linings—only now there were standard linings fitted; of that though, more later.

"How many revs do you want me to use?" I asked Vic. "Ninety-five, maybe nine-eight if you're really trying", he said. Strewth—I see what you get from desmo operation. "Watch the brakes, though. They are only standard linings in the front". The thing that worried me though was the left side kick-starter. But Vic had a dodge to aid starting: a notch cut in the twistgrip rubber that one aligned with a mark on the drum. With these two so positioned a mere dab on the starter brought instant response from the motor, but not unnaturally it seemed a bit odd at first.

When I rode the desmo out onto the track for the first time it was with Clive keeping station behind; no doubt he wanted to see how the opposition was going to shape up. There's something quite different about using a more than four-speed box as one normally finds on a heavyweight. And the box was one that encouraged cogswapping too. Just a dab on the stick and the engagement was entirely positive. I hate to think what would happen if you missed a gear at "ten". But then being a desmo—probably nothing.

Anyway, Clive and I had a pre-arranged scheme to find out how his old Mk I and Vic's new Mach III compared. Side by side down the Norwich Straight we screwed it on together and his valve spring model went into a lead because I wanted to see if I could catch him from behind, and the chase was on. As the rev-counter needle swept around the dial, so the power strokes under the tank gradually merged into a continuous buzz. Higher still until, by the time I slipped into the next ratio with nine thousand indicated, the engine note sounded like a twin and all the usual feel of a single had disappeared; the higher the revs went the smoother was the power delivery. And the back end of the Mk I was much closer.

I caught Clive so fast coming up his slipstream that I was able to sit up and cog up as I went by—I reckon around five miles per hour faster. End of the straight and plant the anchors on and remembering Vic's warning, a little early perhaps. Although the action was maybe a little on the spongy side the retardation was there all right—and I could hear another Duc. not far behind; Clive was still keeping an eye on things.

On now to the quick left-right through the Esses and here the 250 was so easy to throw about and to drive through the corners that it was apparent that I'd have to be going very much quicker to do any good. The technique on a lightweight is so different. You could rush into a corner on the overrun so quick by comparison with, for instance, the Trident, that it was just not true. I could mentally picture Clive rubbing his hands. Later, he said that's just what he was doing. Coram Curve was a piece of cake, though. Nine thou, in third, hard against the stop and she went round as steady as a rock with a feel of the back end just beginning to drift out a little. Through Russell's left/right curves and the rear end started to move a little, past the pits and get set for Riches.

As has been said several times before, Riches has a testing series of ripples in it and everything weaves a little—more or less depending on how good a roadholder the model is. Frankly, the Mach III moved, perhaps a little more than I had been expecting, but I thought I'd get used to it. Gradually, the vastly different technique was acquired during practice. Brake later than I'd ever considered before, wang it into bends with the motor humming like a wasp's nest and bang the throttle on hard before the motor has hardly had the time to take any speed off the bike at all. In fact, the idea seemed to be to forget about stopping the thing so much as one

Photographs: JOHN STODDART

does with a bigger bike; almost, in fact, anything but slow it down—just keep the motor cracking—but hard all the time.

Back in he paddock after practice, I told Vic that I'd been getting nearly 9800 in top gear but he did not seem worried; you'd have thought that the motor was indestructable—he thought it was. "Don't let it go above ten", was all he said. He did have a go at the front brake adjustment, though, and managed to stiffen the action up a little. I did think of suggesting a suspension setting adjustment but thought that having got used to it as it was, I'd better stick with it. I might get it altered for the worse—or better, it's true—but it was a known quantity with it as it was.

Well, we'd proved that the new model was significantly faster than the one that it replaces. There were no doubts, in my mind at least, that the desmo operation was worthwhile. All I had to do now was to see whether the better performance of my model could offset Clive's greater knowledge and experience of dicing a 250 model, to say nothing of nearly a foot in height to be tucked away and an extra two stone. I'm not really a lightweight man, I guess.

The last instruction from Vic I had was, "Enjoy yourself", and drawing grid position number eight, I moved out onto the scene of the action. On the front row was Steve Woods on a 250 cc Honda an his electric starter almost ensured him of an instant start, so he'd be one more I'd have to catch. Graham Penny, too, was close by on Fred Well's similar model to Vic's, save that it was on springs, not desmo. Throw in about four more 250s and you could expect the rest of the forty-strong field to be on faster machinery, an experience that I was not used to and it promised to be interesting, to say the least.

Introducing the NEW 1970

DESMODROMIC DUCATI'S

THOROUGHBREDS FROM ITALY

250cc: £341-10-0   350cc: £373-14-0
450cc: £405-19-0

SALES — SERVICE — SPARES

See these machines at our Showrooms or write for details to:

VIC CAMP MOTORCYCLES
131 Queens Road, Walthamstow, London E17

Tel: 01-520 2093

Clive Thompsett, on his 1965 Mach I model which was unable to hold off the challenge of the rapid Mach III.

The Union Jack floated lazily down as, with throttle carefully positioned as per instructions, my left boot punched the kickstart. It seemed as though all forty motors burst into life at once because I couldn't hear the Ducati's and it was not before I found that I couldn't get it in gear that I realised that the motor was revving much too hard for engagement. Take a breath, roll back the grip a little and now we could get on with the race; but most of the field were on the way to the first corner already. Clive Thompsett in fact, was third round Riches and making ground rapidly.

Right! Head down and give it 9½ in every gear and I went boring round Riches almost without rolling it back and that was nearly a mistake. There's a limit even on a "tiddler". Down the straight, and with my chin bouncing painfully on the tank, the rev-counter needle went up to "ten" and I remembered Vic's instructions. But I certainly was pushing and with the need to catch up, must confess to keeping my fingers crossed and letting her "sing". End of the straight and leaving it painfully late—it would be if I couldn't stop it—banged the gearstick down four times as the Ducati screamed into the Hairpin, snaking about and rushing up the inside of about six bigger machines that needed a lot more time to stop.

After the scratch round the Hairpin the Ducati accelerated out alongside a 500 and there was nothing in it until a ½-litre Triumph got really screwed up in third and then the Mach III dropped in for a tow and outbraked the bigger bike to beat it through the Esses. Round Coram, cornering around the outside of a couple more of the slower men and the Duc, was quite steady when laid over until my boot rubbed the deck. Through Russells curves, though, some movement of the back end became apparent; but it's past the pits for the first time and I'm wondering whether this test will have quite such an auspicious ending as some of the others this year.

Rushing into Riches faster than I'd ever done before, I really pushed it to see just if the wriggle at the back would get any worse over the ripples. Wonderful how blase you can get riding a 250—I wouldn't try that with a 650. Well, wriggle it did but the movement did not seem to be getting out of hand. Another run down that 1000 yard straight and this time I used just about all the brakes that there were and supplemented these by whipping through the gears at the same time while the motor sang a high-pitched tune to accompany the protesting squeal from the tyres.

For another couple of laps I extracted just about everything there was to come out of the Ducati and I'd never have believed that a 250 could go like it, even after dicing with it on my 500 on another occasion. Then, ahead, I spied Thompsett mixed up with a bunch of about four 500s. Ah, ha! I'm in with a chance. Next time round and Clive tried to get on the inside of the group as they went into Sear's right-hand bend, but he got cut off while I was in luck and went sliding round the outside to get in front down Norwich.

Halfway down the straight and the 500s went by and the rev-counter needle went quickly past the 9600 rev mark, though now I sat up to hold it back a bit as there was no point in caning it, having got in front. The 500s start to brake and, still flat on the tank, we went sailing by. Thinks: the brakes can't be that bad.

Eventually I made up enough on the bunch of 500 dicers, so that they stopped passing the 250 down the straight bits and got clear. Clive took a while longer to find a way through, which meant my getting a bit of a lead and a breather. However, another Daytona in front proved to be a tougher proposition and he kept rushing by just before a corner to block the line and force the Ducati to adopt some most unorthodox ones to get by, only to be out-accelerated up the next straight. It really was a quick Triumph.

It soon became clear that if only I could get away then he wouldn't catch me before the next corner and I could make enough time to break away. So it meant that I had to force a bit and dive under his elbow a couple of times before he let me go—and away as I'd hoped. Clive was not so lucky though. Without the extra urge of the desmo, he could never quite compete with the Daytona and finished in 17th place while Vic's bike earned me 15th. Right on the line for the ninth time, Peter Butler came by to terminate the fun and reduce my race length to nine laps compared with his ten.

The Mach III had won its class in spite of my bulk and inexperience with lightweights. Even so, it was several miles an hour quicker than the Mk I model and had averaged something like 83/84 mph to win. The front anchor had lacked something but with standard linings under hard racing conditions I guess you couldn't really expect anything else. The road-holding had seemed a bit skittish over the bumpy bits but with time to set the bike up for my weight I'm sure that could be improved as Clive's model is quite good.

So back to the two-stroke v four-stroke controversy. In spite of the fact that the "strokers" seem to get quicker by the day, and in the racing classes seem to have a definite edge, in production racing circles the bangers are far from past it and won't be, I think, for a few more seasons — I'm happy to say.

Well that's the 250. The 450 is now in Vic's shop for the coming season and I can't wait to wrap myself around one of those. But that must be another story.

CYCLE ROAD TEST:

Three Desmo Ducatis: The Mark 3D 250, 350, 450

The desmodromic Ducati was engineered for the average Italian who rides his bike to work but would rather be a roadracer.

• Every Italian believes deep down in his heart that he could be a national-champion roadracer, if he just had the time to go out and race. Since he doesn't, he wants a lightweight street bike that will go like a mad banshee; will straighten out the curves. Ducati gives him that kind of motorcycle. The U.S. version is changed just enough so that the saddle is comfortable and the handlebars allow the rider to sit upright on the bike if he wants to.

For 1970, Ducati has developed the D-3 (desmodromic) series of street machines—a 250, a 350, and a 450 that vary only slightly in detail, dimensions, and weight, so that the rider can select any type of performance that he wants. These machines are sport motorcycles; they will appeal to the rider who is willing to forego some of the luxury features presently available, in order to get outstanding handling, or performance, or both.

The machines that we got for testing were early prototypes that the factory had sent to Berliner Motors for study. They were fitted with tiny, restrictive mufflers designed for use in another country, and the factory was experimenting with an even-hotter camgrind for the 450's. We used the mini mufflers to get the ET's shown in our specifications column, but they are not truly representative, because when we pulled them off and fitted megaphones, we learned how much stomp these bikes can *really* have.

From the side, the D-3 models all look just alike: lean, low, and sleek. Their proportions are right, they look just as good with a rider aboard. The desmo Ducati is brutally simple. You see frame tubes, and lots of airspace around the engine; no unnecessary sheetmetal. For those who think that a motorcycle looks best if it looks like a motorcycle, the Ducati will go over very well.

We are told that in production, the D-3's will be available either with the scrambler gas tank or the street tank. The scrambler tank (not shown here) has simple, graceful lines, but it doesn't have indentations for the knees. The street tank looks too complicated up front when viewed from a distance of a few feet; but when the bike is fifteen feet away, the intricacies seem to disappear, and the tank looks good, blending perfectly with the front forks and the saddle.

The machines that we tested were finished in candy-red and polished alloy, set off by the black frame and fork-tube covers. The chromed bullet-headlight, the polished sports fenders, the exposed and chromed rear-shock springs—all of these combine to make the motorcycle noticeable, without gaudiness. The alloy engine and wheel-hub castings are clean, even, and precise—a Ducati trademark. And as is always the case with Ducati, the paint has been applied with care.

Like many other brands, the desmo Ducati has its own specific starting ritual. When the engine is warm, it will flood and wet the plug if you are at all sloppy about feeding it gas while cranking it. But your dealer will know the drill. The kickstarter lever is mounted on the left side. The internal ratios are right; the engine is easy to spin. Although the 450 is fitted with a compression release, you don't need to use it.

PHOTOGRAPHY: JOHN SENZER

The desmo Ducati is a motorcycle that looks like a motorcycle—you see frame tubes, and lots of airspace around the engine.

We tested the three desmo Ducatis from Long Island to Delaware. At the New York National Speedway, we ran them through the traps. They were geared rather tall, for high-speed cruising; but even so, when we yanked the mufflers and fitted megaphones, the difference was startling. For example, the 350 with muffler did the quarter in a so-so 17.53 seconds. With a meg, the same bike whizzed through in 15.15 seconds, and friends, for a production-type 350cc machine, that's moving out smartly! We gave all three of the new, still-tight machines a real work-out at the strip; in fact, we overdid it with the 450 and warped the clutch a hair. We concluded that the three engines have a tremendous amount of potential punch, and if the production machines come through with the right mufflers fitted, they ought to fly. The cam in the prototype 450 was impressive enough; if Ducati starts using a still-hotter cam in that machine, the results will be wild.

The Delaware countryside has a lot of rolling hills and gently-winding back-country roads. We had those roads all to ourselves (sharing them occasionally with a '57 Plymouth or two). On the street, the three machines—which we swapped back and forth—handled almost exactly alike; the difference was in performance.

The Ducati's single-cylinder engine ha narrow cases; therefore, the frame, th tank, and the footpegs can all be ver narrow too. You can fit yourself mo easily to a well-laid-out narrow moto cycle than you can to the fat bikes, an the result is a feeling of instant conf dence. Equally important, the machine are built low; even a short rider ca easily plant both feet on the pavemer while waiting for a traffic light t change. The Ducatis feel as if they ha been built just for you, and that the weren't something that came out of crate.

The clutch, while not in the one-finge category, is not stiff. The bikes ar

uipped with a heel-toe gearshift lever the right side of the engines. Since consistently forgot to use the heel rt of the lever, the shifting pattern s: up for low, and down to get into e four higher gears. The gearbox shifts th a tiny "clunk", just to let you ow that all of the splined gears went ere they were supposed to go. We ver once missed a shift. And the tch is not at all fussy or grabby; you n't have to worry about stalling out en you want to get the wheels rolling ain at a stoplight. Even better, neutral RIGHT THERE when you want it! The 250 and the 350 that we tested re equipped with cowhorn handlebars; the 450 had the almost-flat European bars. The cowhorn bars are magnificent. They sweep way up and then way down. With them, we could sit in slight slouch, with elbows bent. The grips were placed at such an angle that you can simply lay your hands on them to retain control of the motorcycle. Grabbing a handful of throttle was almost a sensual experience; due to the angle of the bars, you could rotate your wrist (an effortless motion) to open the throttle, instead of bending your wrist downward (an unusually awkward motion) as most bikes require. The throttle-return spring on the Dell'Orto

Continued on page **107**

• One of the big problems that engine designers have had to face for years has been that of controlling the poppet valves at increasingly higher engine speeds. The valve is usually opened by a cam and rocker arm, and closed by strong springs. But at high rpm, springs can permit the valves to float or bounce. Knowing this, many designs have been set forth eliminating valve springs.

One of the early pioneers was an Englishman, A.F. Arnott, who patented desmodromic valve gear in 1910, wherein the cam follower both opened and closed the valve positively. His design never got into production.

The 1914 Delage GP car used desmo valving. In the Delage system, a stirrup-shaped tappet was attached to the valve. Two cams rotated inside the stirrup: one cam opened the valve, and the other cam closed it. But the engineers were afraid to drive the valve right up onto its seat because of manufacturing tolerances, so they fitted a small spring to move the valve the final few thousandths.

The next important step in desmo valving came in 1954, when the Mercedes GP cars reigned supreme. The Mercedes system used two cams per valve. The opening cam depressed a tappet fitted atop the valve stem, and the closing cam operated a rocker arm forked to the stem. The valve was cam-driven to within .003 inch of its seat, and compression pressure in the cylinder seated the valve. There were no closing springs in the final design. Mercedes claimed that the valve-train of this straight-eight engine could be operated easily by hand, which gave them 30 hp more than they could have attained with regular valve-springs.

Many motorcycle manufacturers experimented with desmodromic valving— among them were BMW, Velocette, Norton, and Ducati. The Manx Norton already had a DOHC engine, so engineer Bert Hopwood decided to use the existing center gearwheel to drive the valve-closing camshafts. The opening cams bore directly against the valve stem tappets, while the closing cams operated through short forked rocker arms. Norton dropped its racing program before the desmodromic Manx could get into widespread production.

The most successful of the motorcycle

continued on page **108**

The 250, 350, and 450 Ducatis are almost exactly alike; the rider chooses the engine that performs to meet his needs.

carb was perfect. Although the slide would close when you released the twistgrip, you didn't have to fight the spring to open the throttle.

The short, flat bars on the 450 weren't so comfortable. They pulled you into a semi-tuck; if you sat forward far enough so as to be able to bend your arms a bit, your knees went farther along the tank than the knee indentations, which was awkward.

The Ducati saddle is a distinct improvement over the thinly-padded two-by-four that the rider used to have to contend with on earlier models. The current version is plenty thick and soft, and it is wider. When we first sat on it, we found that parts of our posterior still hung over the edges, but apparently the saddlemaker knew what he was doing, because the saddle gave us no trouble at all while we were riding.

The handling of the motorcycles was outstanding. We never did encounter any really awful roads on which to push the suspension and frame-rigidity to their limits, but on average-to-poor winding roads, the bikes handled like a champ. The suspension absorbed all of the bumps and potholes that we were able to find, without bottoming. Compression and rebound were quick and positive. The geometry was right, and the machines felt stable at speed, even though it was a lightweight. For example, we overrevved the 250 in fourth gear and rode along for awhile at 95 mph on the speedo, and the bike felt more solid than many machines much heavier than the Ducati.

The motorcycle was supplied with a friction-type steering-damper, but we only used it once, when we encountered gusting sidewinds.

You can lean the desmo Ducati into a corner without having to work at it. You can heel the bike over until you're scraping your boot on the asphalt, and you never get that little nagging worry as to whether or not the motorcycle is about to dump you on your can. It is this feeling of confidence inspired by the Ducati in the middle of corners that makes it so popular as a sport cycle.

Engine performance, as you would expect, differed considerably among the three models. The 250 and the 450 had wide powerbands; both engines began to get at the cam at about 4000 rpm and pulled strongly from there up to (and beyond) the redline at 8500 rpm. The 350 was more highly tuned and had a narrower powerband: the power came in at about 6500 rpm. The 350 is tuned as a street dragster; the 250 and the 450 are over-the-road bikes.

All three of the engines set up a bad vibration at 5000 rpm, which peaked at 6000 and disappeared again at 7000. You'd get this multifrequency vibration through the handlebars, the saddle, the gas tank (which has far too stiff a mounting spring), and the footpegs. It was objectionable when we encountered it going up through the gears, but on the road we ran the engine at higher revs—which is what the desmo valving is all about.

The Ducati designers like to pull a lot of power out of a simple, reliable,

Continued on page **108**

---

250 DUCATI MARK 3D	**350 DUCATI MARK 3D**	**450 DUCATI MARK 3D**
Price, suggested retail .... East Coast, POE $789	Price, suggested retail .... East Coast, POE $839	Price, suggested retail .... East Coast, POE $930
Tire, front .... 2.75 in. x 18 in.	Tire, front .... 2.75 in. x 18 in.	Tire, front .... 2.75 in. x 18 in.
rear .... 3.00 in. x 18 in.	rear .... 3.00 in. x 18 in.	rear .... 3.00 in. x 18 in.
Brakes, front .... 7.1 in. x 1.5 in.	Brakes, front .... 7.1 in. x 1.5 in.	Brakes, front .... 7.1 in. x 1.5 in.
rear .... 6.3 in. x 1.5 in.	rear .... 6.3 in. x 1.5 in.	rear .... 6.3 in. x 1.5 in.
Brake swept area .... 63.1 sq. in.	Brake swept area .... 63.1 sq. in.	Brake swept area .... 63.1 sq. in.
Specific brake loading ...... 6.8 lb./sq. in.	Specific brake loading ...... 6.9 lb/sq. in.	Specific brake loading .... 7.0 lb/sq. in.
Engine type ... 4-stroke sohc single-cylinder	Engine type ... 4-stroke sohc single-cylinder	Engine type ... 4-stroke sohc single-cylinder
Bore and stroke .... 2.91 in. x 2.27 in., 74mm x 57.8mm	Bore and stroke .... 2.99 in. x 2.95 in., 76mm x 75mm	Bore and stroke .... 3.38 in. x 2.95 in., 86mm x 75mm
Piston displacement .. 15.17 cu. in., 248.6cc	Piston displacement .... 20.75 cu. in., 340.2cc	Piston displacement .. 26.57 cu. in., 435.7cc
Compression ratio 10:1	Compression ratio 10:1	Compression ratio 9.3:1
Carburetion .... (1) 29mm, Dell'Orto	Carburetion .... (1) 29mm, Dell'Orto	Carburetion .... (1) 29mm, Dell'Orto
Air filtration .... Felt	Air filtration .... Felt	Air filtration .... Felt
Ignition .... Battery/coil	Ignition .... Battery/coil	Ignition .... Battery/coil
Bhp @ rpm .... N/A	Bhp @ rpm .... N/A	Bhp @ rpm .... N/A
Mph/1000 rpm, top gear 10.8	Mph/1000 rpm, top gear 10.9	Mph/1000 rpm, top gear 13.1
Fuel capacity .... 3.4 gal.	Fuel capacity .... 3.4 gal.	Fuel capacity .... 3.4 gal.
Oil capacity .... 5.2 pt.	Oil capacity .... 5.2 pt.	Oil capacity .... 5.2 pt.
Lighting .... 6v, 70 watts	Lighting .... 6 v, 70 watts	Lighting .... 6 v, 70 watts
Battery 6v, 13.5ah	Battery .... 6 v, 13.5ah	Battery .... 6 v, 15ah
Gear ratios, overall .... (1) 16.66 (2) 11.72 (3) 9.15 (4) 7.45 (5) 6.58	Gear ratios, overall .... (1) 16.45 (2) 11.57 (3) 9.03 (4) 7.35 (5) 6.50	Gear ratios, overall ..... (1) 13.75 (2) 9.67 (3) 7.55 (4) 6.15 (5) 5.43
Wheelbase .... 53.2 in.	Wheelbase .... 53.2 in.	Wheelbase .... 53.54 in.
Seat height .... 29.5 in.	Seat height .... 29.5 in.	Seat height .... 29.5 in.
Ground clearance .... 5.1 in.	Ground clearance .... 5.1 in.	Ground clearance .... 5.1 in.
Curb weight .... 292 lb.	Curb weight .... 294 lb.	Curb weight .... 299 lb.
Test weight .... 432 lb.	Test weight .... 434 lb.	Test weight .... 439 lb.
Instruments .... Speedometer, tachometer, odometer	Instruments .... Speedometer, tachometer, odometer	Instruments .... Speedometer, tachometer, odometer
Standing start ¼ mile .... 18.94 seconds	Standing start ¼ mile .... 17.63 seconds	Standing start ¼ mile .... 16.60 seconds

DUCATI *Continued from page* **107**

high-revving single-cylinder engine, and they get it with the desmodromic system. Four cams on a single overhead shaft move followers that push the valves open and then pull them shut again. Huge mousetrap springs ensure that the valves seat at low rpm, when the valve-closing cam-followers aren't slinging the valves very hard. The result of all this is that the engine just loves to rev.

In fact, the desmo engine sounds very happy at the redline, even when it is new and tight. Although the running-in instructions specifically prohibit doing so, we ran the engines at 8500 rpm just above all of the time—except for the 250. We tried to run it at 8500, too, but whenever we weren't looking, it would ease up to 9300 rpm.

Those plagued little mufflers on the prototype bikes restricted the performance of the machines in the higher gears, but the runs with the megaphones at the dragstrip gave us a good idea as to the strong performance that the machines will deliver when fitted with good mufflers.

The five-speed gearboxes have their ratios spaced out nicely, and if you upshift at the redline, you find yourself right in the middle of the powerband in the next gear.

The cable-operated brakes were smooth and efficient, bringing us to controlled stops every time.

The D-3 Ducati intrigues us. We will be fascinated to learn, having ridden the early prototypes, how the production models for 2970 will turn out. Mike Berliner has promised to lend us one of the first 450's that come in, for our own pleasure-riding. We know a road. . . . ⊙

DUCATI *Continued from page* **105**

desmo efforts was the 1955 Ducati 125cc GP bike. Fabio Taglioni designed this single-cylinder engine with desmo valving, which combined the best features of the Norton and Mercedes designs.

In the Ducati, there were four cams and four rocker arms, although the valve-opening rocker arms were simply interposed between the cams and the tappets. In this engine, there were no valve-closing springs; the cams brought the valves to within .012 inch of their seats, and engine compression seated them.

The desmo Ducati 125s were raced hard from 1956 through 1958, and they proved to be sensational. The single-cylinder engine churned out 19 hp at 12,500 rpm, and a later twin developed 22.5 hp at 14,000 rpm. Ducati came within an ace of winning the 1958 125cc world championship. Alberto Gandossi needed only to win the Ulster GP to win the championship for Ducati, and he was in the lead near the end of the race when he spilled.

For a while, it appeared as though Ducati had forgotten about desmodromic valving, but in 1967 they suddenly popped up with 125cc and 250cc desmo singles in Italian national races. These fast works bikes paved the way for the "D" series street machines.

—*Richard C. Renstrom*

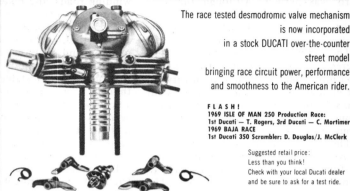

Ducati 450 Desmo

The race tested desmodromic valve mechanism is now incorporated in a stock DUCATI over-the-counter street model bringing race circuit power, performance and smoothness to the American rider.

FLASH!
1969 ISLE OF MAN 250 Production Race:
1st Ducati — T. Rogers, 3rd Ducati — C. Mortimer
1969 BAJA RACE
1st Ducati 350 Scrambler: D. Douglas/J. McClerk

Suggested retail price: Less than you think! Check with your local Ducati dealer and be sure to ask for a test ride.

BERLINER MOTOR CORP.
Hasbrouck Heights, N.J.

CIRCLE NO. 27 ON READER SERVICE PAGE

A TIGER ON WHEELS!

Never has a motorcycle come off the production line with so many features as the new Ducati 350cc OHC Street/Sports Scrambler.

The 1969 Ducati 350 SSS is an entirely new version based on the success of Ducati's numerous national & international competition victories! Wins that proved the stamina, handling qualities and performance of this unique model.

Engineered specifically to surpass the demands of the American rider, the new Ducati 350 SSS is it at home on the highway as well as off the beaten path. Here's a model that is worthy of your attention. See your local Ducati dealer and test ride the '69 Ducati SSS and catch yourself a tiger!

DUCATi
The Thoroughbred of Motorcycles
Berliner Motor Corporation,
Hasbrouck Heights, New Jersey

DUCATI

MOTORCYCLE SCOOTER & THREE-WHEELER MECHANICS

THE ILLUSTRATED HOW-TO-DO-IT MAGAZINE

ENGINE ANALYSIS No. 14

▶ **First home in the 1970 lightweight Production TT! This is a remarkable achievement for a four-stroke single but a fitting result for a marque which has enjoyed a long line of sporting successes.**

The 250 Ducati piloted by Chas Mortimer was prepared by Vic Camp —"works" Ducatis are hardly ever raced outside Italy—and the speed and reliability of the machine is backed up by the fact that two other Ducatis were on the leaderboard. One came in fourth and one eighth.

Undoubtedly, the desmodromic valve gear now used plays a big part in the reliability.

As Mortimer said, "I knew I could over-rev it on the last lap without doing any damage."

A successful ohc single on both road and track

DUCATI ANALYSIS
MAIN SPECIFICATION

Model	250 Monza	250 Mk 3	250 Mk 3D	350 Sebring	350 Mk 3	350 Mk 3D
lubrication Wet sump cast integral with crankcase-gearbox unit. Pressure feed to bearing surfaces from gear pump. Gearbox shafts and gears run in oil bath common with sump						
engine single cylinder four-stroke, ohc with desmodromic valve operation						
capacity, cc	248.6	248.6	248.6	340.2	340.2	340.2
bore × stroke, mm	74 × 57.8	74 × 57.8	74 × 57.8	76 × 75	76 × 75	76 × 75
compression ratio	9:1	10:1	10:1	9.5:1	10:1	10:1
max. power developed at, rpm	7000	7800	9500	8000	8500	7500
rebore oversizes, mm	+ .4, .6, .8, 1.0 →					
fitted ring gap, mm 1st	.25–.40 →			.30–.45 →		
2nd	.30–.45	.25–.40 →		.30–.45 →		
Oil control	.30–.45	.20–.35 →		.25–.40 →		
max. permissible gap, mm	1.00	1.00	1.00	1.00	1.00	1.00
valve timing, ± 5°:						
checking clearance, inlet	.2 mm	.15 mm	.15 mm	.2 mm	.10 mm	.1–.15 mm
exhaust	.2 mm	.30 mm	.15 mm	.2 mm	.10 mm	.1–.15 mm
inlet opens, btdc	20°	62°	70°	20°	65°	70°
inlet closes, abdc	70°	76°	82°	70°	76°	82°
exhaust opens, bbdc	50°	70°	80°	50°	80°	80°
exhaust closes, atdc	30°	48°	65°	30°	50°	65°
running valve clearance, mm	.05–.10	.05–.10	.10–.15	.05–.10	.05–.10	.10–.15
inlet valve lift, mm	7.5	8.55	—	7.5	10.0	—
exhaust valve lift, mm	7.5	8.0	—	7.5	8.5	—
ignition timing:						
static ± 2°, btdc	5°–8° →					
fully advanced, btdc	33°–36° →					
points gap, mm	.3–.4 →					
spark plug	Marelli CW 260N →					
plug gap, mm	.6–.8 →					
bearings: the crankshaft is supported in two roller bearings with roller big-end and plain bush small-end bearings						
transmission gear primary drive to five-speed crossover gearbox via a multiplate clutch. Final drive by chain						
primary reduction	2.5	2.5	2.5	2.11	2.11	2.11
gearbox sprocket, teeth	17	16, 17 or 18	16, 17 or 18	15	14, 16 or 17	15 or 17
rear wheel sprocket, teeth	45	45	45	43	45	40 or 42
internal ratios, 1st	2.46					
2nd	1.73					
3rd	1.34 →					
4th	1.10					
5th	.97					
electrics 6-volt ac/dc lighting and ignition, supplied from flywheel-mounted alternator and battery. Charging current is rectified and voltage-regulated by controlled diodes						
alternator output	6v, 70W →					
battery	6v, 13.5 aH →					
carburettor						
Dell'Orto type	UBF 24BS	SSI 29D	SSI 29D	UBF 24BS	SSI 29D	SSI 29D
choke, mm	24	29	29	24	29	29
main jet	108	112	115	108	112	115
idler jet	45	45	45	40	45	45
valve	80	60	60	70	60	60
atomiser	260	260	260	260A	260	260
float	6.5	14	14	—	14	14
needle	11/2	14/2	14/2	16/2	14/2	14/2

DUCATI ENGINE ANALYSIS

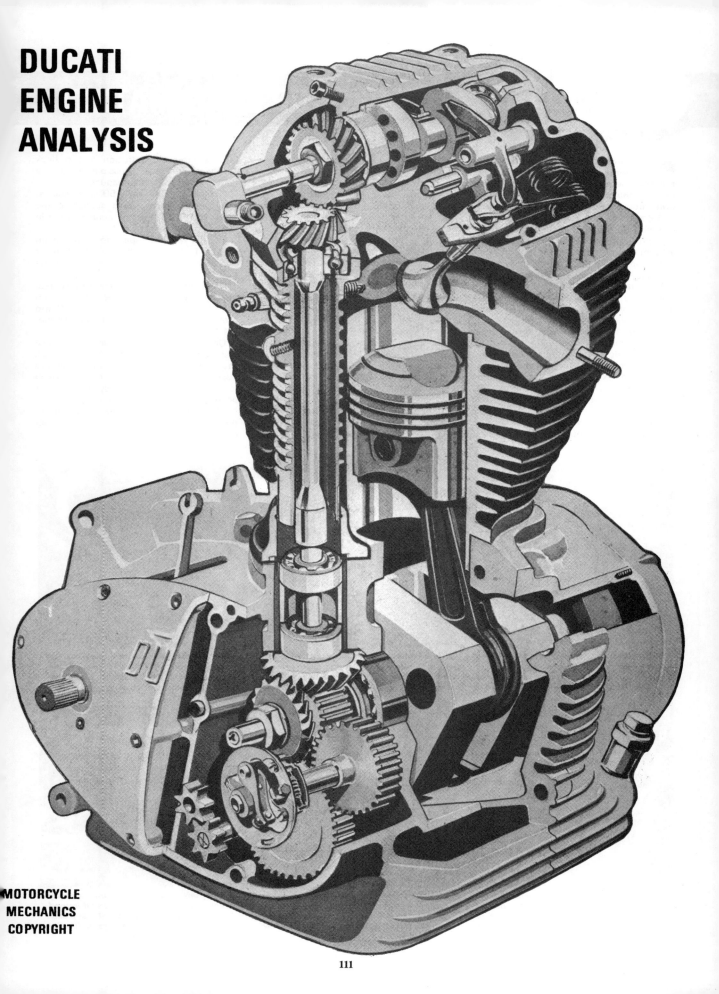

DUCATI ENGINE ANALYSIS
WORKSHOP INFORMATION

▶ The Ducati motors are all essentially the same when it comes to maintenance work. The design is fairly simple, making for easy working and only a couple of special tools are needed.

Because these engines are so simple, it is easy to forget that they are made of high-precision parts which need a certain amount of care and respect.

The specification could be for a racing machine—just remember that!

First, regular and frequent oil changes will easily pay for themselves by the motor's longevity when it is properly looked after.

Second, there are several special tools available which will make life a lot easier if you're stripping the motors frequently, but the ordinary owner only *needs* one. This is an extractor for the alternator rotor, and you risk a lot of expensive damage if you try to remove the rotor without it.

Ducati have now swung right over to desmodromic layout—this has many advantages, like eliminating valve bounce, but the biggest bonus is in reduced wear on the valve gear.

A conventional spring system puts a basic load of 70 or 80 lb. on the valve. This is either taken by the valve seat or acts on the cams and followers. Consequently, the valve seats and opening gear take quite a pounding.

In the desmodromic layout, small hairpin springs are used, just to supply a bit of tension and keep the valves in place, but the loading is only 8–10 lb.

This obviously reduces wear on the valves enormously, but, because they are closed mechanically rather than by a hefty spring, you should pay close attention to the clearances. As there are two rockers to each valve, both the opening and closing are affected by the clearance at the rocker.

When rebuilding the motor, you'll find that nearly all the shafts are shimmed to control end-float. It is a good idea to keep each shim labelled after it has been removed, to avoid confusion, and to check end-float wherever possible with the figures in the workshop manual.

A final note on the 450 cc machines. They are in many ways similar to the 250s and 350s, but, of course, differ in the odd detail.

We did not have space to include them in this feature and the specification for the 250/350 should not be read for the 450.

Workshop manuals, handbooks, spares, machines and advice can be obtained from the U.K. concessionaire, Vic Camp Motorcycles, 131 Queen's Road, London, E.17.

One final point, especially if you have just completed an overhaul, is to check the ignition timing as accurately as you possibly can. Ducatis are not renowned for easy starting and a newly rebuilt motor benefits a lot from precise timing.

Check that the firing point is spot on and the cb points are clean and set to the correct gap.

The head is located by four long bolts. As it is spigoted, it may be a tight fit and need tapping off. Use a rubber mallet. There is no head gasket

The spigoted barrel makes the seal, but a small rubber "O" ring is fitted around this dowelled oilway—renew ring each time the head is taken off barrel

Light hairpin springs are used on the desmodromic valve gear to give a seat pressure of 8–10 psi. Such light loading gives the valves a very long life

The timing gear train is punched to set up alignment—check on the position of the dots before you strip the gears to ensure correct reassembly afterwards

The oil pump drive is taken from the arrowed slot. When replacing timing cover ensure that the oil pump shaft lines up with this slot or damage may be caused

The shafts located in the massive crankcase-sump housing are all shimmed. As all the shims are different, keep each one labelled for correct repositioning

BERT FURNESS OF VIC CAMP MOTORCYCLES SHOWS HOW TO DISMANTLE THE...

DESMO DUCATI

► Italian manufacturers have specialised in speedy lightweights for many years and have gained a reputation for quality workmanship.

The Ducati factory seems to have found the right formula very early because the basic design of their latest Mk 3D remains unchanged except for detail improvements and, of course, desmodromic valve operation.

We went along to Vic Camp Motorcycles, the Ducati specialist, to watch Bert Furness strip the 3D motor.

Like the earlier 250s, this machine was built with performance and low maintenance costs in mind and, given reasonable maintenance, tremendous mileages can be covered between complete engine overhauls. See the complete overhaul overleaf...

Having removed the exhaust pipe, disconnected electrics and clutch cable, the oil should be drained before lifting engine from frame

Four long bolts which hold[s head] and barrel to crankcase m[ust be] removed. Tap head with [a] mallet to loosen. No gasket[s]

Put some insulating tape around kick-starter splines if the unit doesn't need attention. This will save the shims falling off

Undo clutch screws, lift o[ff] cups and springs. Note that [there] is no adjustment—screws m[ust be] tightened fully when rep[laced]

DESMO DUCATI

▶ Although it is a relatively simple motor to strip, the Ducati does need two special tools if damage is to be avoided.

The first and most essential is the rotor removal tool. This is a simple screw-on extractor. Do not be tempted to economise and jar the rotor free with hammer blows—this could prove expensive.

The second tool is a dual purpose one. It can be used on telefork tops as well as clutch centre.

Enterprising Ducati owners can no doubt find an effective way to lock the clutch centre while removing the centre nut, but for the Ducati mechanic this tool will be a time-saver.

Regular and frequent oil changes are a must with all Ducatis, especially if the bike is used for short journeys. With regular oil changes, the motor should not need a complete strip for many thousands of miles. Main bearing wear is negligible.

Desmodromic valve operation has many advantages, the biggest one being the almost complete absence of wear on moving parts and valve seats. Very weak valve springs are fitted to keep slight tension on mechanical parts and so eliminate wear. Valve seat pressure is light—8–10 lb. psi as against 70–80 lb. with the normal valve spring set-up. This explains the absence of valve seat wear—the valves simply do not get hammered.

USEFUL DATA

Ignition timing is set at 34 degrees btdc fully advanced. Points gap—12 thou. Valve clearances—return rocker—nil; opening rocker—5 thou. Crankshaft end float—nil.

On a high-performance motor, where tolerances could be critical, it is important to replace the shims in their right places. It may seem tedious to label every shim, but it is worth it.

Two more points to watch. When removing the selector box cover, you will find that all the holding screws are of different lengths. Put them in their respective holes in the cover or use a suitably marked piece of cardboard.

Secondly, the oil pump spigot (pic. 16) must be lined up with the slot in the pinion shaft when replacing timing chest. Failure to do this will damage the oil pump. If any resistance is felt when fitting timing chest, check these two items at once.

Special tools, spares and even new Ducatis are available from Vic Camp Motorcycles Ltd, of 131 Queen's Road, London, E.17.

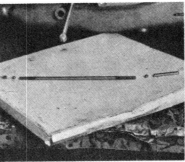

The clutch push-rod assembly must be replaced in the correct order. The short rod (left) bears on pressure plate, ball comes next

The clutch is a very sturdy s[ingle?] plate unit which hardly ever w[ears.] Here you can see the shims (a[rrow-] ed) fitted to kick-starter

Put a bar through small-end eye or against crankshaft to lock motor when undoing mainshaft nut. Triumph rear wheel spanner fits

The gearbox cluster is very [com]pact and sturdy. This five-s[peed] unit should need no attent[ion] providing oil is changed reg[ularly]

the barrel can be removed ecked for wear or scoring. est to keep piston at bottom of its stroke to avoid damage

Prise out gudgeon pin circlip, heat piston and push out gudgeon pin with suitable drift. The pin should slide out smoothly

Check that kick-starter splines are undamaged. The complete assembly is much stronger on later models and should give no trouble

Take off clutch adjusting cover. This will allow a tyre lever to be used to free outer case when the holding screws are removed

move nuts from clutch centre mainshaft, you must first back locking washers. Note haft nut has left-hand thread

Clutch centre must be locked to undo centre nut. Mainshaft nut needs a ring spanner tapped by a mallet to free it if it's tight

Clutch centre and wheel should lift off easily—a sliding fit. A spacer is fitted behind the clutch housing, don't forget it

A special tool is needed to remove rotor. There is a Woodruff key on mainshaft. Use tyre levers to ease pinion nut from shaft

ve rear cover, then points and centre screw. Now the fixing pillars and five ining timing chest screws

Important: when replacing timing cover, the spigot (indicated) must be aligned with its slot. Failure to do this will cause damage

Valve timing is marked by dots. It is as well to check these before you take the pinions out. This timing applies to all models

Note the shim on either side of this pinion. Tape shims on to the shaft to ensure correct replacement. Check pinion teeth for wear

you can see the kick-start hanism and the massive main ng which lasts for very big ages under normal conditions

Every shaft has a shim and all the shims are different. Label them carefully and put them aside to avoid possible wrong replacement

Very light valve springs keep a slight tension on mechanical parts to eliminate wear. Desmo gear will last life of bike—leave it alone!

Although no head gasket is fitted to any model, there is a small rubber "O" ring which must be renewed every time head is lifted

THE DUCATI THAT RON BUILT

A 391-cc Dirt Special That Is Almost Too Pretty To Ride

WHEN RON WOOD, of Costa Mesa, Calif., decided to build a "play bike," he certainly wasn't "playing." Beginning with an old Ducati 250 engine he had around the house, and a Bultaco Metisse frame that was given to him, Ron started burning the midnight oil and came out with a truly beautiful and unusual creation.

The engine features a 350 Ducati crankshaft and connecting rod assembly which is capped with a piston made from a Forgedtrue blank, machined to give a compression ratio of 9:1. Early oiling problems were finally solved by using a Perfect Circle three-piece oil ring. An increased bore size of 81-mm necessitated making a new sleeve, giving the engine a new capacity of 391-cc or 24.2 cu. in.

Cylinder head work includes enlarging and reshaping the inlet port out to 1 3/16 in. A 1 3/16 in. Amal Monobloc carburetor is now used with a filtron air cleaner element. A larger inlet valve has been fitted and the original hairpin valve springs have been replaced with double coil springs which have titanium collars. A sports grind cam from a 450 Ducati has intake and exhaust lifts of 0.400 in. and 0.385 in. respectively. The exhaust port has also been enlarged and a 1 5/8 in. diameter header pipe 21 in. long fits into a short, shallow-taper megaphone, which helps maintain good low-end pulling power.

Aluminum bronze valve seats are used, an additional spark plug is mounted at an almost horizontal position just behind the cam drive tower, and two additional oil drains emerge from the valve pockets on the left-hand side of the head.

Thirty bhp at 8750 rpm were achieved at the rear wheel on C.R. Axtell's dynometer recently, with good power coming on from 6500 rpm. It's certainly no trials bike, but it compares very favorably with 360 two-stroke motocross machines.

Ignition is accomplished using 12-V electrics. Honda coils are fed by the original alternator which was rewound to get the necessary 12-V current. Ignition timing is now set at 39 degrees BTC with no automatic advance.

Below, all the gears and the clutch wheel were drilled for lightness. A more recent five-speed scrambles gearbox helps keep the revs up and the power coming.

But, however interesting the engine modifications may be, the frame claims just as much of the spotlight. Ron first sandblasted and thoroughly cleaned the frame, inspecting everywhere for cracks or broken welds. Then he set about the *coup de grace:* a rubber-mounted engine! Using Norton Commando "Isolastic" suspension components, he cleverly grafted them into the Bultaco frame. By changing the thickness of the spacers, the engine can be more or less rigidly mounted to achieve the desired results. Some experimentation was necessary, but the machine is now quite free of vibration, which has always been a problem with medium- and large-displacement, single-cylinder machines.

After the engine suspension components were installed, Ron filled all the welds with a plastic filler, ground them smooth and painted the frame with a special, high-gloss black paint. Front suspension is Ceriani. A 175 Ducati front hub is laced to a 21-in. alloy rim which sports a 2.75-21 Dunlop Trials Universal. The rear hub is laced to an 18-in. alloy rim, and a 4.60-18 Dunlop Sports knobby is fitted. The only other frame modification was the repositioning of the rear shock absorbers to clear the chain. Weight with a half-tank of fuel is 256 lb.

A homemade seat graces the topside and a coat of "Kawasaki racing green" (also called competition yellow) finishes one of the prettiest "play bikes" we have ever seen.

(ABOVE) The Norton Commando "Isolastic" front engine mount. Note the rubber-mounted exhaust pipe and additional rocker box drains. (BELOW) All welds are filled with plastic filler and ground smooth. The "Isolastic" suspension unit looks like it was designed for this machine.

CYCLE WORLD
ROAD TEST

IS THERE ANYBODY out there who appreciates the virtues of a 500 class four-stroke Single any more? (Echo.) Hmmm, not many left, maybe. Too bad people have to be so damn serious about specialized dirt machinery, light weight and all that stuff.

For those of us who still appreciate the vibrant throb of a Roto-Rooter, its single power pulse traction, and torque from practically zero rpm, there are only two machines. The BSA 500 and the Desmo Ducati 450 R/T. BSA offers its Single in three models: motocross, dual purpose and roadster. The Italian Ducati goes a slightly different route, offering you a sparse basic package with a lighting kit, then leaving you to puzzle out how to make the package fit your needs.

The 450 has that neither-here-nor-there quality of non-specialization, the one that says, "Set me up the way you like me." As delivered, it works best as a woods bike, but begs modification for desert, fire road riding, maybe even Sportsman TT racing. Given a loving and clever owner, a Roto-Rooter can be made to do anything. And it sounds and feels so good!

The Ducati R/T is not dual purpose in the strictest sense, as it is not completely street legal in all states when its lighting kit is installed. The kit, consisting of a headlight, taillight, wiring and handlebar switches, does conform, however, to most federal licensing standards. The notable exceptions here are lack of a device to activate the brake light, and lack of horn and muffler. The kit does provide the basics though; all that is needed to complete the package is a little extra work and a few inexpensive parts.

The Ducati R/T's biggest drawing card is its single overhead cam engine with desmodromic valve gear. A desmodromic valve system is one that mechanically opens as well as closes the valves, as opposed to the conventional system in which valves are mechanically opened by a camshaft/lifter arrangement and then closed by a strong spring. Desmo valves have been considered advantageous because valve floating at high rpm is completely eliminated, thus insuring greater reliability. Another less obvious advantage is that less internal friction is created in the valve mechanism, yielding increased power output over a similarly tuned engine with a conventional valve train.

While use of a desmodromic valve system is quite rare, it is certainly not a new idea, and it has been proven to be quite reliable in competition. In 1955, designer Fabio Taglioni completed his first desmodromic engine, a 125-cc Single. A year later, Degli Antoni piloted this 125 to victory in the Swedish Grand Prix. What's more, only a crash at the Ulster GP the following year kept the Desmo 125 from defeating MV for the 125cc world title.

In the racing version, valves were closed mechanically to within 0.012 in. of their seats. It was left to inertia and compression pressure to seat the valves. The engine worked super at high rpm, but idled erratically, due to improper seating of the valves. Kick starting was impossible because the engine had little or no compression at cranking speed.

Perhaps this is why Ducati did not use the desmodromic valve principle on their production bikes until 1968. The 450 Desmo is a more simple design than the racer was, but, like the racer, it is completely reliable. In the 450, a single overhead cam with four lobes is driven by a bevel gear on the right side of the engine. Followers mechanically open and close the valves, which are now seated by springs. Hence, the R/T starts easily and runs well at low rpm.

Carburetion is handled by a 29-mm Dellorto concentric and, as expected, the power band is very broad. In fact, the 450 pulls well from a little over idle to its 7000-rpm power peak. The 38 bhp produced is delivered to a close-ratio, five-speed transmission by helical cut primary gears. Shifting is positive. Internal ratios are well spaced, but overall gearing is too low to take advantage of the engine's broad powerband.

An ignition switch is not fitted and the engine is stopped by a compression release. The release is mounted in the side on the cylinder barrel just behind the camshaft bevel gear, but it sticks out too far to be protected by the engine. If the bike is laid down on the right side during a crash, there is a strong possibility that the release and its cable will be damaged. Fitting a longer cable and rerouting it behind the frame tube just to the rear of the release would help.

As a trail mount, the Ducati R/T is superb. It is particularly suitable for Northern California or East Coast terrain where trees, gullies, and mud bogs keep the speed down. Steering is quick and light, allowing excellent control at slow speeds. Bars, too, are comfortable when the rider is standing, and they are sufficiently wide for good leverage. Ground clearance is a generous 9 in., and rocks and logs can be easily cleared. And, while the ultra low gearing is useless on the street, or in high speed, off-road situations, it is handy when riding in close quarters. Quite simply, the R/T is a climber. Hills present no problem—either going up or coming down. Brakes are seldom needed, due to the engine's good braking characteristics, and, of course, the bike's low gearing.

Unfortunately, all of these characteristics (low gearing, high

DUCATI 450 R/T

Meet The Desmo-Rooter:
A Set-It-Up-The-Way-
You-Like-It Charger.

DUCATI 450 R/T

ground clearance, quick steering) which make the R/T excel in the woods, work against the bike for high speed, off-road riding.

Aided by a quarter-turn throttle and low gearing, the engine reacts quite rapidly and doesn't seem to possess much flywheel. Wheelspin is excessive—much like a powerful Twin. For high speed work, the bars are too high (the R/T also has a high steering head) and are swept back, forcing the rider to the rear on a machine that has a light front end. The R/T slides easily, but these slides are difficult to control, unless you sit forward against the fuel tank. Steering is ultra rapid. On relatively straight sections of fire road or in sand washes, the R/T twitches from side to side, sometimes to an alarming degree if the surface is irregular.

High speed stability and slower steering, however, can be achieved by either modifying the triple clamps or the frame itself to increase front fork rake. Both methods have proven successful.

While frame geometry is completely different from the 450 Desmo Roadster, design is similar in that the engine is used as a structural member of the frame in both applications. Three toptubes are incorporated in the unusual design. The center toptube passes from the top of the steering head all the way back to the upper shock absorber mount. At this point, a hoop to support the rear fender is bolted on. The single downtube terminates at, and is bolted to, the engine. Two parallel tubes which pass downward to the swinging arm pivot from the aft toptube crossbrace support the engine at the rear. The swinging arm passes inboard of the frame tubes, as is common practice. The footpegs, made of 3/4-in. OD bar stock with a plate welded on top, are of the folding, spring-loaded variety.

In an effort to make the riding characteristics as adjustable as possible, four upper shock mounting points are provided instead of just one. The position most rearward, which makes the shocks practically vertical, raises the rear of the bike slightly and transfers weight forward. Moving the upper mount forward angles the shocks more. This gives the swinging arm increased leverage and softens the ride.

A friction type steering damper which passes through the steering head is particularly useful on the Ducati, as the steering is extremely quick. A fork stop prevents the forks from damaging the gas tank. The unit consists of a tab welded to the front downtube which contacts stops on the triple clamp. It's not nearly strong enough, though, and one crash will bend it considerably.

Of interest is the rear wheel adjusting mechanism. It consists of a cam-type washer shaped similar to a French curve. The axle hole is offset. When the washer is turned counterclockwise, it rotates against a tab welded to the lower shock mount, and moves the wheel farther back in the swinging arm. The washer has been reference-notched to allow easy visual alignment of the rear wheel.

The front forks are another new item. Manufactured by Marzocchi in Italy, they are of Ceriani design and have 7 in. of travel. The spring rate is well suited to the R/T Ducati and damping is variable by substituting oils of varying viscosity. The ride is very soft, certainly a plus factor on long rides in the boonies.

The rear shock absorbers don't work nearly as well. When the rear brake was applied on downhills, the wheel hopped, indicating that they have inadequate rebound damping. Substituting a set of Konis cured this completely.

Both front and rear brakes are full width units, quite large for an off-road machine. Lever pressure is light, but tends to be a bit grabby and causes the wheels to lock up easily especially on downhills. These units should be easier to use however, when the linings become worn a bit.

Knobby patterned tires, a 4.00-18 at the rear and a 3.00-21 in front, are mounted on Borrani units. They are of Akront design, but unlike Akront rims, they bend easily. Rim locks, a must on a torquer like the Ducati, are also absent. Two locks should be installed in the rear rim and one in the front to prevent spinning a tube.

Controls are first rate. The levers are soft, have a good feel, and bend rather than break. Rubber covers connect the cable housings to the lever assemblies and, like Magura levers, there is a large knurled wheel that allows the cables to be adjusted while the bike is being ridden. A handlebar-mounted choke is bolted through the right lever. The compression release on the left can be reached while still holding onto the handlebar grip.

The R/T's body components and stylized fenders are made of fiberglass, but not of the flexible variety. Consequently, the fenders break all too easily, even though they are rubber-mounted. The high-mounted front fender presents no clearance problems, but the same cannot be said of the rear unit, which isn't wide enough to cover the tire and rubs frequently. Ducati dealers recommend substituting yellow Preston Petty unbreakable fenders for the stock components; this would solve both the clearance and breakage problems.

Another potential problem is the fuel tank. After 100 miles or so of trail riding, stress fractures appeared in the gel coat. These were probably caused by vibration, and careful mounting of the tank may or may not cure this. Two fuel taps should provide good fuel delivery, even when the tank is low. The petcocks, however, are mounted up in between the frame tubes quite close to the head. They are difficult to get to when the engine is cold, and are impossible to reach without burning a hand when the engine is warm. While shutting off the fuel taps isn't usually necessary on a four-stroke, it would have been better to mount the taps elsewhere.

Unlike most Italian bikes, the seat is soft, wide, and provides good support for the rider. It's long enough for packing double too; all that's needed are a set of passenger pegs.

Fiberglass side panel/number plates conceal the aircleaner, which is centrally mounted just in front of the rear fender. The 8.5-in. diameter by 2.5-in.-thick unit houses a paper element that does a good job of filtering as long as it is dry. The only thing wrong with this aircleaner is that it is difficult to service in its present location. The backing plate is held on by a single wingnut which is easily removed, but the bottle for the chain oiler and the close proximity of a portion of the rear fender mount make it impossible to get the cover off without removing the bottom mounting bolt and bending the aircleaner forward slightly. At best, it is both difficult and inconvenient.

The chain oiler consists of a plastic bottle, with a capacity of approximately a pint, mounted to a frame tube just above the chain. A valve at the bottom of the bottle controls the flow of oil. The valve can be shut off completely if oil is not desired, or the rate of flow can be adjusted as necessary.

For a new design, the R/T Desmo Ducati shows much potential. As delivered, its soft ride and broad powerband will endear it to trail riders who spend long hours in the saddle. Add a speedometer and spark arrester, then "dial in" the handling, and the R/T should be competitive in enduros as well. The possibilities are endless, for this is a machine that can be easily modified for satisfying results in a variety of uses. ◉

DUCATI 450 R/T

SPECIFICATIONS
List price	$1189
Suspension, front	telescopic fork
Suspension, rear	swinging arm
Tire, front	3.00-21
Tire, rear	4.00-18
Engine, type	four-stroke, sohc Single
Bore x stroke, in., mm	3.38 x 2.95, 86 x 75
Piston displacement, cu. in., cc	26.57, 435.7
Compression ratio	9.0:1
Claimed bhp @ rpm	38 @ 6500
Claimed torque @ rpm, lb.-ft.	N.A.
Piston speed (@ rpm), ft./min.	3195 @ 6500
Carburetion	29-mm Dellorto concentric
Ignition	flywheel magneto
Oil system	gear pump, wet sump
Oil capacity, pt.	5.28
Fuel capacity, U.S. gal	2.25
Recommended fuel	premium
Starting system	kick, folding crank
Air filtration	dry paper

POWER TRANSMISSION
Clutch	multi-disc, wet
Primary drive	gear
Final drive	single-row chain
Gear ratios, overall:1	
5th	7.86
4th	8.91
3rd	10.94
2nd	14.01
1st	19.95

DIMENSIONS
Wheelbase, in.	56.5
Seat height, in.	33.0
Seat width, in.	9.5
Handlebar width, in.	34.0
Footpeg height, in.	14.0
Ground clearance, in.	9.0
Curb weight (w/half-tank fuel), lb.	285
Weight bias, front/rear, percent	44.2/55.8

DUCATI 250 24 HOURS

DUCATI are well-known as makers of motorcycles with two outstanding qualities. The engines are robust, precision-made; and their handling is second to none. All Ducatis have these virtues, I think, and none more to than the latest model to be imported into this country, a two-fifty named the 24 Hours model. The 24 Hours differs in one way from previous Ducatis sold here in that it is made under licence in Spain. To the paying customer, this is important because he gets a very much cheaper motorcycle when it is made in Spain than he does when the same model is produced in its native Italy. The natural question is: What about quality then? This could only be answered satisfactorily if one were able to stand the Spanish Ducati side by side with the Italian version and compere specific points. As the equivalent model is not made in Italy we were not able to do this. We are bound to say, though, that Spanish quality is not *quite* as good as we have become used to from Ducati over the years. The engine is every bit as good but the paintwork did not seem to have such a deep lustre to it. As it is considered that this version could sell for up to £100 more if it were made in Italy, prospective customers may well consider that tips the balance.

Have no illusions about the Ducati 24 Hours. As its name implies, it is in essence, a road-going production racer, and the designer had this in mind when he sorted out its priorities. Comfort did not figure very high on his list and the prospect of carrying a pillion passenger not at all; for the firm dualseat, very much in the Italian tradition, allowed the rider a bare 21in and the riding position insisted that he used every inch. The whole concept of the Ducati's riding position is based upon its being used on the track. Clip-ons, a long narrow petrol tank and footrests which if not rear-sets are at least well back allow one little choice of riding position. It is very neat and stylish but, at least until one's neck muscles become used to it, extremely uncomfortable. It becomes less so if one really plays racers and adopts a full-scale racing crouch, but almost anywhere on public roads this is liable to attract unwelcome attention. We settled for a semi crouch, back at about 45 degrees to the horizontal. For the first 400 or so miles a stiff neck resulted but, as we say, the muscles soon adapted and it became bearable. We would not be so happy at the prospect of a 2,000-mile continental tour though.

One has to look very hard indeed these days to find a decent-size four-stroke single; in fact apart from the BSA 500 we cannot think of any outside of Italy (we are no more able to think of Ducati as being Spanish than we are of Aermacchi being American). It is a near tragedy for the present-day newcomer to motorcycling that he will probably never get the chance to ride a four-stroke single. The concept is not perfect but it does have a charm and character all of its own and, on the rare occasions that we get one for road-test, the memory of its thump-thump-thump exhaust note and deceiving power lives long after the machine has been returned. In a way the 250 Ducati is not really in the British tradition of a single. It is too small for the diehard big-single man and it goes in for

bevel-driven overhead camshafts, a refinement which our designers in the old days usually saved for full-blooded racing motorcycles. In the case of the 24 Hours the designer has been particularly shrewd about this. He has put a glass cover on the bevel housing on top of the engine so anyone who wants can watch the busy little bevel gear doing its job and at the same time observe the effectiveness of the oiling system. Who can resist such a fascinating sight? No I.

The Ducati engine is the sort that makes friend and converts wherever it goes. It must be over 10 years since we were astonished by claims that a 250 Ducati in production trim could exceed 100 m.p.h. They are still at it! This tiny little 250, weighing in at only 260lb, reaches a claimed 106 m.p.h. Our efforts to substantiate this will be the subject of a later test for our machine was almost brand new and we were unwilling to ill treat it so early in its life. Suffice to say that our machine had no difficulty in reeaching 90 m.p.h. without being flat out. That extra 16 m.p.h. should not be too difficult to find. How does it do it? The 247 c.c. engine, with a bore of 69mm and stroke of 66m, is as well put together as a watch. The compression ratio is 10 to 1, high but not unduly so. The o.h.c.-driven valves are inclined at 80 degrees. The Amal carburettor, sucking great gulps of unfiltered air, has a 28mm choke. The crankshaft is supported by two main ball-bearings with a smaller bearing on the flywheel alternator side. Unusually for a continental machine, the five-speed gearbox has its change on the right side but keeps its kick starter on the left. The footrest had to be folded up before the starter could be used . . . a chore.

Ducati know a thing or two about brakes and those fitted to the 24 Hours were as good as they needed to be. Our test model had a single-leading-shoe front brake; a good one too. Production models will be fitted with a twin-leading-shoe version, which should make good braking even better. A wire spring is fitted to the outside of the rear-brake drum, presumably to ensure a full return quickly when it is released.

Ceriani, who seem to be making forks for almost every bike we try these days, are responsible for the excellent unit on the Ducati. It is very much more robust than is usually found on a 250.

24 Hours is a title that implies that the machine is going to be used, has been used or will be used in a 24-hour race . . . something like the race of the town where this machine is made, the Barcelona 24 Hours, run on the twisting Montjuich Park circuit. To achieve this one needs lights, very much better lights than are fitted to the machine we used. These lights were not really good enough for medium-speed road use.

Sitting on the Ducati, one has the thought that it doesn't really feel like a 250. The narrow petrol tank and tight, neat, riding position give the impression that this is something less than a 250. The kickstarter provides the first inkling one has that this is not a run-of-the-mill lightweight. To start one has to flood, generously. No air control is fitted. A hefty, determined swing of the kickstarter always produced results. Anything less was liable to bring a sharp reminder that this engine had a 10 to 1 compression ratio and sporting valve timing.

The long, black megaphone-type silencer produced a most attractive noise, a dull, almost flat, boom that increased in volume as the revolutions rose but never, somehow, offensively. At least, not to me. Give me the thump of a four-stroke any day. Using the Ducati about town is not really what it is made for but any road-test bike we have has to earn its keeep and it performed its chores surprisingly willingly. Perhaps its only vice in this respect was a tendency to expire with a splutter if one opened the throttle too suddenly. Then came the chore of folding a footrest up before restarting.

It came as rather a surprise to find that a tachometer is not fitted to the machine as standard, the test machine having a k.p.h. speedometer only. A tachometer is available as an optional extra but one would seem to be almost statutory if the machine is going to be raced.

Our question, 'Cafe racer or real one?", was encouraged by the fact of the Ducati making such a first-rate "café racer". It is forgiving, glamorous, fast and makes the right kind of noise. It could be used for such a shallow task, but that would be a pity. The open road is the place where the 24 Hours began to show us its talents. The faster one went and the more one pushed the Ducati the better it responded. Few singles make a big drama of going fast and the 250 o.h.c. engine never seemed to be making hard work of giving its best. By the same token many singles have a vibration period. The Ducati's was at 60 m.p.h., showing itself with a drumming through the glass-fibre tank and a tingle through the handlebars. It was soon over and never gave acute discomfort.

Even the open road is not the real happy hunting ground of this sleek little red production racer. This is the race track, showing other 250s that the day of the four-stroke single is not yet done. Vic Camp, Ducati concessionaire, who supplied us with the test machine, is entering one in the 250 class of the production TT, to be ridden by his very successful runner, Alan Dunscombe. As a follow-up to this story we have been offered a ride on this machine on the track. Perhaps then we may be better able to judge the Ducati 24 Hours for what it is. Meantime we have found that it is a more than acceptable road machine from a performance point of view but its quality as a long distance road-riding machine is impaired by the "racer" riding position. It is in this country at £352.32, which makes it a very serious competitor to most other performance 250s on the market. We look forward to completing this test at Brands Hatch for then we have a feeling this machine is going to come into its own.

B.P.

WRAGGS of CHESTERFIELD

Stockists for

DUCATI

250 c.c. NOW IN STOCK

SERVICE & SPARES

95 Lordsmill Street, Chesterfield.

Tel.: 3622 or 78540

• I discovered it in the May 1959 issue of *Cycle*, and called it the Ducati with the funny front brake. In two months I had nearly worn out Page Ten, eyeballing the photograph of that motorcycle. And now it was June 1959, the middle of final exam week at Northwestern. I was holed up in my sterile dormitory room with its beige walls, beige bunks and blonde woodwork, resolutely entrenched behind stacks of textbooks, piles of notebooks, and a forty-pound heap of dirty laundry. My head throbbed and bulged with more German declensions, more themes of Western Civilization, more geologic formations, and more notions about American poets than anyone should ever care to know. My head was packed with 270 facts-per-square-centimeter; the addition of just one more detail would have broken my jaw or fractured my skull. Except for one thing. At its innermost vital center, my mind was cheating; it kept focusing on the Ducati with the funny front brake.

Right there on Page Ten sat that bike with Franco Farne, an Italian racing champion. Farne had won the Class 4 lightweight race at Daytona with his 175cc Ducati: a low, lean, hard little machine with an enormous double-scoop front brake. The motorcycle was so purposeful, so elegant, so perfect. Raw, green lust streaked my desire for a machine like that.

Pavement racing, I decided at the time, was the ultimate form of motorcycle competition. I reached this conclusion, not because I knew anything about motorcycle competition, but because I had had vast experience with a Ford tractor and all the dirt-work that people do with such devices—plowing, disking, harrowing, cultivating and

COLOR PHOTOGRAPHY: JIM McGUIRE

SATISFIED MIND

BY PHIL SCHILLING

Franco Farne and his 175 Ducati Formula 3 locked together at Thompson Raceway, 1959.

planting. My early years spent aboard a factory-Ford four-cylinder colored for some time my view of *any* off-road activity—and my judgment of engine technology: aluminum alloy I loved because cast iron I hated; overhead camshafts I applauded because flatheads were an anathema to me; I worshipped beautiful forks and shocks because that Ford tractor lurched and jumped like a hiccup-crazed drunkard as it yanked a plow through rock-infested fields. The Ducati with the funny front brake shared nothing with that damn Ford tractor save the theory of internal combustion.

Reality kept intruding. In June 1959 the magic little Ducati lay almost half a continent away from me, and the doorless, windowless, hatchless omnipresence of Northwestern's final week surrounded me. My summer was likewise sealed; it held neither time nor money to go chasing after some funny little motorcycle. I was reduced to all that I could be: 1959's World Champion Magazine Racer.

Back in those dim days, it took real talent to be a magazine racer. In the first place, you needed a good magnifying glass for close inspection of magazine photographs which were—with frustrating consistency—small, fuzzy, underexposed, overexposed, poorly cropped, and frequently miscaptioned. Furthermore, any magazine racer worth the price of a year's subscription had to have an uncommonly creative imagination. Lightweight roadracing reports were either brief or non-existent; at times only the finishing orders found their way into magazines. Depending upon how thin or thick the shreds of information might be, you could divine your own race reports. But lots of ordinary gruntwork and big rolls of postage stamps were still necessary for a proper magazine racer. You wrote for literature, for "additional information," and sometimes you turned joiner.

I wanted to follow the exploits of the funny little racers, so I joined the Worldwide Cycle Club, which was yet another promotion of wheeling-and-dealing, dodging-and-darting Floyd Clymer, who during that era published *Cycle*. I did not join the WCC (Fastest Growing Cycle Club In The World!) for my "serial numbered membership certificate suitable for framing for your room, office or den," nor for the "beautiful 20K gold two-color pin with screw-on post lock for lapel or cap," nor for the "pocket secretary with notebook, ballpoint pen, calendar and card holder with attractive club emblem on the front cover." No sir, Floyd. I joined because you said that local papers would be mailed to members the day after race events, and those events included Daytona, Laconia, Dodge City and, as an extra attraction, the Isle of Man. And Floyd, though you passed on some years back now, I'd just like to say that one P. Schilling, WCC Member #2347, never got his papers. So I never found out anything through the WCC about the little Ducati racers. Floyd, it was a bummer.

I scoured each new issue of *Cycle* for more and better photos of the machines with the funny front brakes, and was always disappointed. I wrote for free literature from Berliner Motor Corporation, the Ducati importer, and asked about the 175s with the special brakes; in the mail, five days later, came a four-page brochure on the new Ducati street motorcycles which I added to my collection of identical four-page brochures. So I wrote again, and asked about the funny-braked bikes again, and into my dorm mailbox five days later there dropped three goodies. The first piece was yet another four-page brochure; the second, a newsletter which had a picture of the 175 OHC bike with the funny brakes; third, a note which said the bikes with the funny brakes were basically just Ducatis with a few special light touches here and there.

I zeroed in on the evidence at hand with all the zeal and energy of a Methodist missionary. I calculated, guesstimated, cross-checked, double-checked, and almost burned out my left eyeball. But in the end I congratulated myself on my sleuthery: the 175s with the funny brakes had been lightly touched all right, touched again and again and all over; though similar to the street machines, they were *real racers*, genuine single-purpose machines. I confess that much time could have been saved either by going to a race (an impossible feat at the time), or simply by calling the distributor and asking point-blank about the racing machines. But in 1959 I shunned long-distance telephone calls; dormitory pay-phones gulped down nickels, dimes and quarters with much eager clattering and greedy dinging. Besides, a five-minute call would have deprived me of two months of intriguing diversion and daydreaming. Long-distance directness seemed far too businesslike for a World Champion Magazine Racer.

For Joe and Mike Berliner, who were launching the Ducati name in the United States, racing was a serious business. Only a small number of Americans had ever heard of the name, and only a handful had ever seen a Ducati. If one had followed the European racing scene, then he knew that the Ducati triple-cam desmodromic 125 single had just missed winning a World's Championship in 1958, that the diminutive Bologna racers owned a string of victories in national

Ron Dahler: a racer who remembered everything.

events in England and Europe, that Fabio Taglioni (who designed Ducatis) was regarded as a genius, and that Franco Farne had been three times a Junior Champion in Italy. These things Joe and Mike Berliner knew, but they realized something even more important: to most American motorcycle enthusiasts of the Fifties, what happened in Europe was just about as significant as who won the last scrambles race of 1958 in Wabash, Indiana. As Walter von Schonfeld (Berliner's public relations man and race manager) pointed out, if you wanted to claim that Ducati motorcycles were fast, reliable bikes which could be hurled around corners and pulled to a stop quickly, then you had to prove these claims, and prove them on American ground. Which is exactly what the Berliners did. And that explains the bikes with the funny brakes.

They were genuine production racers; the official designation was Formula 3. The Ducati racers bore as much resemblance to their street counterparts as the present-day Yamaha TR3 roadracers do to Yamaha 350 street machines. It depends upon your point-of-view: you can argue that such racers are apparently similar but fundamentally quite different, or apparently different but fundamentally quite similar. In the end, the argument turns on what part you're looking at, and what importance you attach to differences. The Formula 3 departed from the 175 Super Sport street bike in many ways. The most distinctive feature of the racer was the brakes. The front unit was a 180mm four-shoe affair with single leading shoes on both sides. Two giant intake scoops dominated the front brake, while the 160mm rear brake had a rear-facing air intake. The Formula 3 frame was lighter, lower and longer than the standard model, and the racer carried different forks and shocks. The racer and road bike didn't share the same frame geometry either. Neither the Formula 3 steel tank nor racing saddle could be bolted on the street machine. The engine castings of the Formula 3 were sand-cast, the street bike engine die-cast; the castings themselves were all similar, but not exactly the same. Camshafts, rocker arms, connecting rods, pistons, flywheels, bearings, clutches, and transmission gears all differed from racer to road machine. Primary drive gears in the Formula 3 were straight-cut, and so were the bevel gears; the standard bike had much quieter, but less efficient, helical-cut primary and bevel gears. Formula 3 racers weren't built on any normal production line; they were put together individually and very carefully.

Of course the general layout and design of the Formula 3 followed the street pattern, and when you uncrated your racer, you found that it had full fenders, muffler, an electrical system, headlight, taillight and license plate holder. But you sensed that Taglioni designed the Formula 3 for number plates, not license plates. All those light touches began at the drawing board.

Before long, Americans with 175 (and later 200) street Ducatis were doing some touching of their own. They discovered that the standard models could be modified, that "speed parts" were available for track-bound road bikes, and that if you had enough horsepower and skill, you could live without sand-cast engine cases and fancy brakes. The days of the woodshed tuner had not yet passed; the power which 200cc Triumph Cubs made certainly was proof of that. Yet those good old days were numbered, and years wore numbers too, and Tom Compton knew it.

By 1962 Tom Compton had been roadracing for four years. His battle-scarred Parilla showed its age: quick-fixing in the pits on race weekends, falling behind out of corners, and slipping farther back down long straightaways. Compton had both the money and racing record necessary to buy a Formula 3. Money alone wasn't usually sufficient for at least two reasons. First, there were very few Formula 3 racers available; the entire factory production of 175 racers totalled about fifty. Second, by 1962 Berliner no longer had an official, centralized racing program, so they didn't want to waste good machines on hopeful beginners. Farne had raced in the United States in 1959; Francesco Villa succeeded Farne as the distributor hired gun in 1960, and like Farne, Villa reeled off a series of wins. But when he packed off to Italy at the end of 1960, Berliner's race-to-race, day-to-day involvement ended. Joe and Mike Berliner had successfully launched the Ducati name in the United States, and racing had served its intended purpose.

Ducati racing became privateer racing, and leading privateers kept the Bologna machines competitive: Ron Dahler, Chuck Andrews, Ray Hempstead, Jim Hayes. Roadracing was small, clubby and friendly. When Tom Compton wanted to know exactly how much more carburetion the 175 could take, how the intake port should be shaped, what number needle, slide and jet, what gearing, and a host of other whats, Dahler (who had a super-quick 175) provided the answers, parts and some extra tips. None of "Well, you might try . . ." or "I've never done anything to mine . . ." or "Can't help you, bud." Rather it was "Do this, and then do this . . ." and "Here, you'll need this tool." Guys like Dahler made the sport.

Compton raced on through 1962, 1963, 1964. He loved his 175 Formula 3; it was an Italian single, a factory job, a sophisticated piece of hardware. On occasion, the bike spit him off, but that was racing; at other times it stopped, like at Daytona '64 (with a chipped valve), but nothing runs forever. Old racers don't die, they just get tired. Or outclassed.

Lightweight roadracing was growing up. 175s grew to 200s. Then 250s replaced 200s. Increasingly, 175s had to run with 250s, and anybody's good 250 would just blast down a first-rate 175. Only in small-bore classes did a 175 have a chance. So Tom Compton decided to unload his 175 at the end of 1964. The cam was worn out, and the rockers with it; the connecting rod needed replacement. But Chuck Beer wanted the bike for under-200 racing. Compton sold it.

Time worked against Beer. He had to tear the engine down, locate all the needed parts, rebuild the engine, and maintain one or two other racing bikes at the same time—and pursue his Ph.D. in Physical Chemistry. Rebuilding the Ducati became an impossible, strung-out project. The engine went with Beer to State College, Pennsylvania; he would never reassemble it. The bike stayed in Detroit, Michigan, carefully stashed away in the basement of his father's house. And there it waited for five years.

Meanwhile, 1959's World Champion Magazine Racer had been on the move. I collected my degree from Northwestern and moved on to the University of Wisconsin, where as a graduate student I was getting pretty good at peering solemnly into the grain of seminar tables and pondering outloud that "we should be careful to differentiate the nature of the circumstances from the circumstances themselves." Now I never really understood what I meant by such a phrase, but I knew why I used it. Driving a 1500-mile round trip over a four-day race weekend didn't leave a lot of time to go lurking around inside the Historical Society Archives or the University Library. When you hit town six hours before the seminar meets, you had to fake it.

By the end of 1965, I had hung around roadracing long enough to learn some important things: a four-day diet of peanut-butter sandwiches and coffee won't kill you after all; you can drive 80 mph on the Queen's highway in Canada and get away with it; New Jersey hates speeders; trackside outdoor johns are dangerous to health, especially the breathing function; always pool tollway money for both ways before the trip begins; never turn down a ride-along passenger (no matter how crowded it gets) so long as he's bringing real cash money.

I learned other things at the race track. Notice the word is *at*, not on. I was always the guy sitting on the pit wall, a stopwatch in one hand and a wrench in the other. I admitted more might be learned from behind a tach than a stopwatch, but fast road riding had already taught me one hard lesson: everytime I started to go really fast, the machine just suddenly disappeared, and I was left bounding along in twelve-foot 90 mph strides. I usually managed the first two steps, but the third one always got me. Moreover, I had this vision of sitting there in Tuesday seminar in my body cast, and whistling through toothless, swollen gums: "Uhh thee wee mahzee deefeeuhheeeahh een ahh naarurr ah ahh rrrcummanncees unn ahh rrrcummannees eemmells."

By 1966 I developed a very cold, heady, stopwatch outlook toward lightweight racing. Any fool who cared to look could see that no one would ever make a competitive single-cylinder four-stroke again, or even a single-cylinder two-stroke. Two-stroke twins had everything—if you let sentimentality get in your way, you dumbly ignored reality. Stopwatches do not lie.

Nelson Ledges, 1966. I remember looking at someone's ratty, grunchy old 175 Formula 3. It appeared as if it had been tossed down the road twenty-dozen times, hammered and screwdrivered apart, cobbled and pasted and wired back together. It looked sick, it sounded sick, it was sick. Sand-cast engine. Fine! But low-number horsepower. Straight-cut gears inside. Super! But so what. Close-ratio four-speed gearbox. Neat—only about two gears shy. Lovely brakes. Great! But just about fifty percent overweight. A genuine pedigreed racer. How nice! You can wave its proper papers at those Gyt-Kitted Yamaha 100s come streaking past. Technological progress, I reminded myself, cannot afford compassion; such an old Formula bike was fit rubbish for racing' garbage cans. My mind mocked that relic with all the cold, unfeeling precision of a 17 jeweled stopwatch. My heart, however, believed not a word. It was a bike with the funny brakes, and the magic was still there.

So it came to pass in November 1969 that I climbed down a set of cellar steps in Detroit, looked to my left, and saw the ex-Tom Compton Ducati 175 F3 resting in a corner of Chuck Beer's basement. To me, the bike seemed just as purposeful and elegant as Farne's machine on Page Ten. But it wasn't quite as perfect. As a matter of fact, there was a large gaping hole where the engine

Golden wings, green garland and a big red D.

was supposed to be, since Chuck still had the dissected engine in Pennsylvania. The little Ducati had been parked there for five years, and now it was going away. But I wasn't buying it. My friend, Howard Sprengel, wrote the check.

Howard and I had a plan. Though the scheme didn't fall into that euphoric "British World-Beater" category, our master plan did have a few elements in which ambition overreached good common sense. We knew this, but on the whole, we had decided after plotting for months, our scheme was at least plausible. We told each other that Howard

(Opposite Page) No Italian racer of the period is complete without two things: a Dell'Orto carburetor and lots of brass safety wire. (Above) Simple, classic beauty is a trademark of the single cylinder racer.

(Right) The sand-cast single-overhead camshaft engine (62mm x 57.8mm) produced maximum power at 9800 rpm and drove the tiny Bologna racer just past the 100 mph mark.

was a Canadian Junior class roadracer, that you really didn't need a Yamaha to produce competitive lap times in the Junior class, and that a fully-modified Ducati 250 engine, when slipped into the 175 running gear, would make a sensible alternative. In any event, we always added, the 175/250 would be reliable.

Our grand theory got into trouble almost immediately. First, Howard was promoted to the Senior roadracing class even before the Canadian season began. Second, the trick 250 engine did strange things to the handling of the 175 racing frame. If Howard really stuffed the hybrid into fast corners, he found himself atop the world's most sophisticated and powerful (but dead reliable) OHC pogo-stick. Since neither Howard nor I was very enthusiastic about breaking Howard or bending the pogo-stick, we packed off. After an exchange of parts and pieces, the kind known only in motorcycle circles and in marble-swapping sessions of ten-year-old boys, the 175 was mine.

By this juncture, a large wooden strongbox had arrived from Chuck Beer, who had oiled, wrapped and carefully packed every part of the 175 engine. When he had disassembled the thing in 1964, Beer made notes, diagrams and drawings as he went along. His scholarly, scientific mind had not allowed him to do otherwise. I was delighted. Beer had not hammered-and-chiselled away at the engine; there were no chips, no rounded edges, no cobblemarks anywhere. And Beer had gathered most of the new parts necessary to rebuild the engine.

Understand that there were mysteries. The engine has a kind of arrogance about it. There are no reference marks on individual gears in the valve and ignition geartrains. There are no index dots marking the proper keyway on a couple of gears, and you have your choice of three keyways on each gear. You are expected to know. How much preload should there be on the angular thrust main bearings? How do you set up the cam timing perfectly when the infinitely adjustable exhaust lobe can be locked into place anywhere on the shaft? You are expected to know. In a dozen different ways the engine yells at you: What, you dummy, you don't know? Then what is a klutz like you doing trying to assemble an engine like me? What lovely arrogance.

But I had the mysteries licked, because I knew what to do. Call Ron Dahler. He had the answers.

I had to know, I had to do it right. Because the factory worker who built the engine in the first place *cared*. He cared enough to make everything fit perfectly, enough to match every set of gears exactly, enough to drill lightening holes in every possible gear. He might have hidden a sloppy detail deep inside the engine, but he didn't. He could have shrugged off an almost-perfect fit, but he refused. He didn't shave time; he spent it, lavishly. Maybe he enjoyed standing around at odd moments with the world's most perforated clutch drum in hand, tapping the edge and listening to the clear ring. I sure did.

If you look at a piece of metal in cold, functional, computor-like terms, then it just doesn't matter whether or not it rings true. Yet I see things differently. There's more than metal in that clutch drum; there's a piece of someone's life in there. And when it rings, it's just the emotion coming out.

One August evening in 1970 my little bike with the funny brakes stood completed. Our motorcycling clan gathered on my back lawn as dusk faded toward darkness. I was proud, tanked full of confidence. I had brought the 175 back to one piece, and would now demonstrate before the gathered multitude that the machine would start and run. Eleven years after I eyeballed Page Ten, nine years after Tom Compton bought it, six years after Chuck Beer disassembled it, and one month after I littered the living room carpet with it, I shoved the bike forward, dropped the clutch and listened to the engine gurgle, cough and then break into a loping, free-spinning racket. The sound waves coming out the megaphone had sharp edges all over them. I was ecstatic. We wheeled the little racer into the garage, closed the door, warmed the thing up, and all went deaf. Since there was no chance of running the bike up and down the street, we adjourned to the kitchen refrigerator and broke out a case of wine.

Revelry followed. I must rely upon my friends for a full accounting of that evening. They claim to have last seen me passed out in the bedroom wearing this satisfied smile. I claim to remember nothing.

Two months later, I left for New York to join *Cycle*, and early in 1971 the 175 took its place in the *Cycle* Magazine Shop—that huge, air-conditioned mecca filled with machine tools, one-of-a-kind motorcycles, and rows of tool chests. The shop gave me this feeling of happy fearlessness about my 175. What a relief not to worry about breaking anything on the machine; hell, I cheerfully counseled myself, you just about could make any part in the world here. The perfect set-up. But not quite.

My apartment phone rang early in the morning on Friday, June 24, way too early for good news. The very best thing it could be was a wrong number. It wasn't. Toni Thomas, Jess' wife, reported in calm, measured tones that there had been a fire at the *Cycle* Magazine Shop, and a number of bikes had been roasted, and my 175 looked, well, really bad, maybe completely destroyed. I went numb. Somehow, it seemed to me, I should catapult out of bed, land on the floor, beat my fists on my chest, and howl out a Charlie Brown AAUGH! But I did what I always do at the peak of a crisis: go numb, drop back twenty yards in my mind, pull four bulletproof glass walls squarely around the crisis, and wait for the numbness to wear off. Surely, I argued with myself at a distance from the problem, disasters like this just don't happen to big-league motorcycle journalists. There should be rules against such things. There ought to be good reasons why crazy damn accidents don't plague journalists like myself. Well, I couldn't think of any.

Later, when I could pick my day, I saw the 175, and it was awful. The little bike squatted there in the burned-out section of the shop, like an immolated waif in a black-rimmed crater. Every surface in the shop lay frosted with soft, black ash. The worst thing was my prize: tires burned and charred, rims buckled and distorted, paint blistered and peeled away. Limp glass fibers outlined the shape of the accessory tank and seat. Soot coated the front brake, but the rear brake hub was a heart-smasher—in the heat of the fire, some spokes had pulled right out through the hub flange. Where once there had been plastic-covered cables, now there were only rusty coils. One shock absorber fractured at the upper mounting boss. The green plastic cover which had been thrown over the bike melted into mung that filled every odd space in the sooty engine castings. Every spare part for the machine had been destroyed. I had but one consolation. The fire missed the original tank and seat.

Take the insurance money and run, advised the office wags, and buy some other piece of old Ducati exotica. I couldn't. I could not toss out my 175. A neat desmo, a special one-off racer or anything else did not relate to Page Ten, and Page Ten was the lodestone. Besides, the burned-out 175 was *my* 175. I found the classified ad for it, drove a thousand miles to get it, swapped bikes around to possess it, suffered through the agony of rebuilding it, and shared in celebration of those who designed and built it. That bike was locked up with eleven years of longing. The numbness was gone, the bulletproof glass walls fell away. Neither time nor money nor frustration nor discouragement counted. Only a fierce, self-determined will to re-create the bike mattered.

Cycle cannot give sabbaticals for emotional fulfillment. The editors fuss with personal projects on their own time, and those projects bear the editors' own gruntmarks. For seven months, every spare evening, every holiday, every odd Saturday, I walked the six blocks between my apartment building and Cook's loft, where slowly and relentlessly I rebuilt my 175. The resurrection project became a time-study operation. I discovered how much time is required to peel, scrape and brush that sticky, plasticized mung off engine cases (twelve hours), how long it takes to burn my hands in a straight solution of Janitor In A Drum (three hours), and how many sessions are necessary to sand fire-cured paint off of pitted frame tubes (37 sixty-minute sets). And I learned other things. Cook Neilson instinctively understands the ring of the clutch drum; my spouse grasps the special relationship between the little Ducati racer and her husband, and my dog is still jealous of all motorcycles.

If the first restoration had presented intriguing problems, the re-restoration only gave me pounding headaches—you can't find much romance in five slimy gallons of Janitor In A Drum. Every straightforward task developed crooks and kinks. Redoing the original gas tank was typical. The tank had leaked around one petcock boss, and that leak had kept both the tank and original saddle off the bike and out of the fire. For that I was thankful. Once silver soldering cured the leak, then the tank could be handed over to the painter. Sorry, not so fast. Rust had been breeding inside the tank. Not just little patches here or there, the rust held every square inch of territory inside the tank. I explored all kinds of methods for driving the rust out—and tried more than one.

Shaking a tank with fifteen pounds of pebbles ricocheting around inside is fun for the first twenty minutes. When the twenty-first minute arrives, you know you're not having fun anymore. Your ears begin to rattle, and your arms and shoulders report pain. So plug the earphones into the stereo system, set up four hours of hard rock on the changer, and dial the volume up to seven. Stand in the middle of the living room, wired into the music, bugalooing the tank in double time. Hope the music will block out the pain which is spreading from your shoulders to your back. Be glad you don't need to explain what's happening to your dog.

The snags didn't end with tank cleaning. I wondered if there would be any original decals to put on the tank after it was repainted. Locating three of the four proved simple, but the fourth transfer I never found. The angst over the decal was nothing compared to my anger when I discovered that the chromeplaters lost the neckbolt for the gas tank's air breather. Did they find it? No. Could I easily run down another fitting? No again.

So that was the tank. And for every hangup the tank gave me, other components presented about three more: frame, swing arm, forks, shocks, brakes, hubs, sprockets, tachometer and controls.

I skinned knuckles, tore hangnails, waited out delays, chased after parts, went down dead ends, stopped, backed up and restarted again. How, I asked myself, could this be fun? It just wasn't. The first restoration was fun; it had been a delightful crusade. The second restoration began—and continued—as a grim obsession hardened by a resentment toward the fire and capricious reality. There was no fun in turning reality around; there could only be the satisfaction of having done it. Inside my head I carried this dream-like vision of the completed bike. I knew exactly how the sand-cast engine cases would look, I could see how the rich colors would work together, I could visualize how all the brass safety wire would be wound. I could focus on every spoke, every little fitting, every hose, cable, wire, nut and nipple. I vowed to push, shove, lever and beat reality until it corresponded to my vision. And in March 1972 reality and vision merged.

The bike is as it was in the beginning. So purposeful, so elegant, so perfect. The magic is still there, and my mind is content.

●

People, Places & Parts

A restoration project depends upon three things: (a) basic and detailed information about the motorcycle, (b) original parts, (c) people who can rebuild or duplicate original parts, or who have abilities outside the competence of the restorer. And the novice restorer (the category into which I fall) should understand that patience and money are essential prerequisites for the successful completion of any Projectus Magnamus. If you have a small bankroll and a low threshold of frustration, do yourself and your psychiatrist a big favor: forget restoration and go collect bottlecaps.

Basic Information: No American magazine ever published a track test of a Ducati 175 F3, and there aren't any old advertising brochures lying around. There may be some piece in British or Italian motorcycle journals, but I was unable to uncover that information. Since the bike's previous owners preserved the factory specification sheet and tuning data, necessity never required transoceanic scrounging for information. The spec sheet is in Italian, but finding an English-Italian dictionary was simple compared to any and all possible alternatives.

Sometimes distributors can provide basic information about deceased motorcycles. Bear in mind, however, that distributors are in the business of selling new motorcycles and parts, and they don't have unlimited time to devote to someone's special problem. Furthermore, personnel changes occur normally over the years in any business. Consequently, if your interest is lodged in a 1932-whatever, you're not likely to find someone (under the distributor's roof) who remembers everything about your long-extinct prize. Indeed, you might discover all the ex-importers are firmly grounded in six feet of earth and feeding daisies.

Fortunately, 1959 wasn't *that* long ago. Joe and Mike Berliner, who were the movers and shakers of Ducati racing in the United States, remain. So does Walter von Schonfeld who ran the racing program for Berliner Motor Corporation. And Reno Leoni, who is head of Ducati service at BMC, knows an enormous amount about Ducati racing bikes. They understand Ducati nuts.

People whose lives at one time became entwined with a particular brand

PEOPLE

of machine are usually the best sources for information and parts. If your interest lies with vintage Popes or classic Indians, then you begin by ferreting out an owners' or collectors' club. If, however, you're rebuilding a twelve-year-old racer, then you must find someone who raced or owned a similar model, someone who hasn't tossed out all his clippings and old photos—and if you're really lucky, someone who has kept his old parts.

Tom Compton and Chuck Beer (the first two owners of my bike) kept almost all their correspondence relating to the 175, so I found the right ex-racer—Ron Dahler. Dahler was invaluable. He had a 175 in 1960-61 which was a jet. When the 250 Ducati roadracers superseded the smaller models, Dahler sold his 175 to Oreste Berta in Argentina, a fact which suggests how desirable that particular 175 was. Dahler remembers everything about the early racers: which cams with what timing worked the best, what clearances should be run and how critical they were, what things to do and those to avoid—*everything*.

Parts: Jon White, at Sports Motors in Cincinnati, Ohio just might be the chief Ducati F3 parts-maven this side of Italy. Bob Schanz, the head of Sports Motors, has been a keeper of assorted Ducati exotica for years, so the enthusiasm runs in the company. Jon came up with a number of old special pieces which Berliner Motor Corporation (the major parts source) could not supply.

Cosmopolitan Motors in Hatboro, Pennsylvania still has a vast store of things Italian, and some important pieces which were common both to Ducati and Parilla racers came out of the Cosmo parts bins. Replacement alloy rims posed no big problem, and North American Wheel in Anaheim, California turned up a set of replacement spokes.

Hands: Most of the sweatwork I did. Some things I can do, and other things I cannot, and I credit myself for knowing the difference. When you're working with a part which, to the best of your knowledge, is the last available one of its kind, you can't chance blowing it. In the Great *Cycle* Magazine Shop Fire, some spokes pulled directly through the flanges of the magnesium-aluminum rear hub. I thought it was a goner. No one, but *no one*, happened to have an identical rear hub lying around, though Jon White came pretty close. There was only one thing to do: get the original hub fixed. The task wasn't made any easier by the fact that nobody knew just what alloy-mix the hub contained. It was exactly that sort of situation in which you must know some expert firsthand, and have absolute confidence in his ability. I could trust Tim Horn in Madison, Wisconsin with the mind-snapping job, so I

Continued on page 158

BY PHIL SCHILLING

Some forty kilometers separate the Ducati factory and Imola on the map. A short distance. But it's a damn hard trip when you're going to Imola to win.

Road To Imola

Cav. Fredmano Spairani, Direttore Responsabile Del Coordinamento E Della Produzione, led the way, clicking along with swift, precise steps. Tuesday evening, April 18: four full days stood between Ducati and the Imola 200, and Spairani, the chief administrative officer at Ducati, ignored the gruelling day behind him as we walked toward the racing shop. I was trying to visualize what the racing version of the Ducati V-twin would look like. Certainly, I reasoned, the bikes would not be very similar to the street 750s, because even the most innocent observer knows that superstrong AMA-style 750 racers begin with standard engine cases—but there the kinship ends. So, as I tailed Spairani through the door to the racing department, I expected to see two or three very, very special Ducati 750s.

The doorway opened into a huge shop where about a dozen mechanics were working away on eight motorcycles. The first thing I saw, the thing that immediately dented my mind, was a centerstand. These factory racers were all parked on centerstands, *stock* centerstands, which were connected to stock frames, which joined standard front forks and near-stock swingarms. And the production-line frames held embarrassingly standard-looking engines. Sure, there were special pieces: big Dell'Orto carburetors, high-rise/low-rise megaphones, dual discs in the front and single discs in the rear, oil coolers, hydraulic steering dampers, and racing shocks. But where were all the *really* trick parts? The one-off frames? Special forks? Jet-fill gas tanks? Electronic ignitions? Titanium clip-ons? Wind-tunnel designed fairings? Where were the air-cooled clutches and dozens of odd little sand-casted parts? There weren't any. The machines just looked too standard to be proper racers. Fabio Taglioni knew better.

For seventeen years Taglioni has designed every Ducati worth remembering. He is a reflective, soft-spoken person whose far-

PHOTOGRAPHY: CHEH NAM LOW

Mention Ducati in Italy and two names almost pop up automatically: Bruno Spaggiari, evergreen racer at 39, and Dr. Fabio Taglioni, the chief designer—quiet, reflective brilliance.

ranging interests mark him as a modern-day Renaissance man. He has done many things successfully, but he never embellishes his achievements with the rococo of braggadocio. He is one of those rare people who can wield a cigarette-holder with absolute ease and not a trace of affectation. He offers thoughtful answers, makes cautious understatements and gives studied responses. Those around Taglioni pay him the ultimate compliment: when he says something, they listen, carefully.

Taglioni came to Daytona this year, investigating AMA 750 racing, measuring the level of competition, and calculating the time and resources necessary to produce a competitive Ducati Formula 750 racer. He discovered an army of Japanese racers at Daytona, machines which in his judgment were well-conceived and developed. In Taglioni's view, the Japanese solve their technical problems through a mathematical approach because the Japanese can afford to do things that way. They work by the numbers; it's efficient, logical, straightforward. If machinery is short on horsepower, cylinders are added. And since Japanese companies have scores of technicians and computers at their bidding, they can organize task forces, promptly grind out data, and develop new designs on short order. Italy is not Japan. Ducati could not afford to play the numbers game for a 750 racer.

Taglioni is *the* designer, and he is only one man. He cannot subdivide a racing project into scores of separate parts and assign problem areas to individual teams. He has no computer; he jokes that he doesn't even know how to use one. So Taglioni relies on

Modena sorting-out day: Spaggiari hurls through a 100-plus mph sweeper (Top Left). *Smart froglegs out of a left-hander at Modena* (Top Right). (Near Right) *Twin-ignition, two-valve head; camdrive towershaft is at right.*
(Far Right) *Stock frame, forks, cases, centerstand; there were no really trick pieces.*

134

The desmo rocker set-up. Loop-spring on the forked exhaust rocker insures valve seating for starting.

90-degree twin breathes in through 40mm Dell'Ortos, produces 84 horsepower at 8800 rpm.

what he calls his intuition, a sixth sense which has been sharpened and honed by experience—and quiet reflection on that experience. It's a feeling for what will work, and work well.

The 750 racers are very new, and yet very old. Judged by the mainstream standards of multi-cylinder design, the Ducati twin is dated; yet the engine is sophisticated in its tributaries. Taglioni did not guess his way to success with the 750 racer. Much of the technology which went into the big twin came from the single-cylinder 350 Desmo racers which the factory produced in the late 1960s. And some lessons learned from the 500cc twin-cylinder racers (1970-71) could be applied to the larger twin. Although the first 750 racers were built and run briefly in 1971, serious work on the racing 750s began shortly after Taglioni returned from Daytona.

Taglioni wanted a totally balanced machine. Ducati could never match peak power figures of the multis, but winning dyno-contests wasn't all important. The object was to design a total machine in which handling and braking were matched to usable horsepower. Taglioni got the kind of power he was after. According to the factory, the Imola engines produce 84 horsepower at 8800 rpm at the rear wheel, and at 7000 rpm the engine delivers 70 horsepower.

The source of this power is in the cylinder heads of the 750 racer. Each cylinder (80mm x 74.4mm) draws fuel/air mixture through newly-designed 40mm concentric Dell'Orto carburetors. The carbs sit on the end of five-inch manifolds, and the cylinders breathe in through 40mm valves and blow out through 36mm exhaust valves. The valves are opened and closed by a four-lobe single overhead camshaft. In this desmodromic system, the valves are opened by conventional rocker arms, and a set of exhaust rocker arms close the valves in lieu of coil springs. Each exhaust rocker wears a loop spring; the springs are only effective at low rpm—they simply insure that the valves seal properly on the bronze seats for starting. At running speeds on the desmo racers, the exhaust rockers (and cylinder pressure) seat the valves. In the 1950s, when valve-spring technology lagged behind engine speeds, Taglioni was one of the pioneers of desmo valve control. Valve springs have been developed a good deal since the 1950s (and Ducati has built valve-spring 750 racers), but Taglioni prefers the desmo system despite the incredibly close tolerances required. In Taglioni's judgment, the desmo layout still gives more accurate and reliable valve control, especially when dealing with relatively large, heavy valves. And the desmo system can withstand more abuse of the redline than valve-spring models. The 750 Desmo redline stands at 9200 rpm.

The 750 racers run a compression ratio of 10:1 which is a point higher than the street models. The domed, three-ring slipper pistons form considerable squish areas at TDC with the combustion chambers. Since the racers have relatively high compression engines, oil coolers are fitted in order to hold down operating temperatures. Moreover, twin ignition makes for cooler running. The main purpose of the dual-plug heads is to cut the advance back to 34 degrees BTDC, which in turn lowers the cylinder head temperatures.

Ducati forsook electronic ignition on the racing 750s because the heat generated behind the fairings resulted in some component failures, so the racers reverted to a straight "total loss" battery-and-coil system with contact breaker points.

The lower end assembly of the racer is

LONG ROAD

stock—except for the connecting rods which are machined out of forged billet. They are stronger and 50 grams lighter than standard units. But everything else downstairs in the 90-degree twin has been picked off the production line. There is nothing special about the gearbox; there are no changes in the internal ratios. Even the primary gears and clutch are standard, though the clutch drum has been drilled in order to save weight.

The 750 racing engines bolt into standard frames. Of course, there are extra little tabs welded on here and there, pieces which hold on fairing mounts and foot pegs. Nevertheless, the frames are standard in both geometry and materials. Head angle is the standard 29 degrees, and the front forks which have been taken off the assembly line, give about four inches of trail. There are two standard discs (280mm x 7mm) upfront rather than the standard machine's single disc on the left side. (The stock model carries the mounting bosses on the right fork leg for an extra disc.) The rear disc (230mm) is something special—the road bike has a drum brake on back. The caliper units are Lockheed, just like the street machines. Total dry weight with fiberglass saddle, tank and fairing is 392 pounds. The fairing isn't super aerodynamic, but the racer will pull the tallest possible gearing which produces 169 mph.

Taglioni believes that further increases in power will require building a lighter, stronger frame; he does not expect that a re-framed racer will be a lot smaller physically than the present machine. In any case, a further developed 750 racer will also be a "total machine." More power can be pulled out of the 750 racing unit, but just exactly how much more, Taglioni doesn't know, and idle speculation doesn't fit his style. There's a bit more horsepower, he says, and perhaps there will be experimentation with four-valve desmo heads. But Taglioni is not a man to worry about the last horsepower at the last rpm. Lap times at specific courses, with specific riders concern Taglioni. He's been there already, and seen it all.

Wednesday, April 12: It's another chilly, wet morning in Bologna. The heavy clouds aren't threatening, they're making good on the threat. Outside the rear door of Ducati's racing department, Franco Farne, the head wrench, bumps off each Ducati 750 racer before loading the bike into the big glass display van. Wednesday is not an Imola practice day; it is a Modena sorting-out day, a hasty session to acquaint riders with machines and to decide who shall ride at Imola. Farne brings one big twin after another to life, and everyone is trying to decide just what the 750s sound like. It's the first time the Imola bikes have been started.

Paul Smart is huddled in the courtyard with Vic Camp, the British Ducati importer, and Alan Dunscombe, Vic's rider. Taglioni joins the huddle with Bruno de Prato, Ducati's press officer. They talk about Paul's on-again/off-again Imola. Smart originally planned to ride a Triumph three at Imola, but then the political maneuvering for his British triple began, and Smart was off his Triumph. So Imola was off for Smart. Then calls, messages and wires from Camp who set up the Ducati deal at the very last minute. Smart flew to Italy. What the hell, it was a ride, wasn't it? He could collect some start money, and take his chances along with everybody else. And you never really know about these new bikes, do you? The question was: if it kept drizzling, could anyone find out anything about the 750s at the Modena airport circuit? And Smart wanted to know.

Dunscombe, who has ridden Camp-Ducati singles in club events in England, is waiting for his first crack at a big Ducati racer. Back in England, he and Camp have been playing with 750 street machines, so Alan was not getting aboard a completely alien machine. Even so, Alan knew, the factory racer would be different, and Imola could be Alan's big opportunity. He came to Italy to make the Imola starting team. And he would.

So would Ermanno Giuliano, a tiny Italian rider who is dwarfed by the 750 racer. Giuliano, though little known outside Italy, spent the 1971 season on the 500cc Ducati twin, and had a couple of rides on the first tentative 750 racer built in 1971.

Bruno Spaggiari, along with Smart, was an automatic pick for the starting line-up. Mention Ducati in Italy, and two names immediately pop up: Taglioni, Spaggiari. Spaggiari is 39 years old. He has raced for Ducati in the 1950s, through the 1960s, and into the 1970s. He has ridden singles, twins, fours, single-cammers, double-knockers, single-cam desmos, triple-cam desmos, 125s, 175s, 200s, 250s, 350s, 450s, 500s, 750s. What Ducati has built Spaggiari has raced. When the machines were competitive, he could win. When the machines were outdated, he kept trying. When the bikes went slower, Spaggiari tried just that much harder. He never quit. Spaggiari, evergreen at 39: the years have made him wise, but they have not slowed him down.

At Modena: the clouds broke up over the circuit, and the flat, fast, bumpy roadway dried out. Stopwatches were ticking—following the progress of Spaggiari, Smart, Giuliano, Dunscombe and four lesser Italian lights. The riders were probing, trying to find out where all the boundaries were; trying to process and plug in all the information coming back from bars, pegs and saddles; trying to observe, experiment and check again. In the pits Spairani, the administrative architect of Ducati's corporate resurgence, watched in his intense, electric way. It was the first time the Imola racers had been run, and it was three days before the race. If there were any congenital defects anywhere, the time for correction had long passed. Taglioni was there too, serene and cool. Spaggiari was going like a jet. Smart, circulating slowly at first, changed tires and then reeled off a series of quick laps around the airport circuit.

Spaggiari continued to put in the fastest laps. The veteran Italian appeared to have the best handling bike, though it was difficult to be certain on that point. Spaggiari was on home territory, and he was riding the kind of machinery which has been almost part of him. He was expected to be quick, tidy, precise. He was. Slicing through a sweeper at far beyond 100 mph, Spaggiari just folded away under the windscreen, and every adjustment he made came out in fluid movement. His bike did not perceptibly wiggle, side-step or judder through the sweeper. The machine simply tracked through the corner, the suspension taking up the bumps and keeping the tires on the ground, the engine sending out a flat, edgy drone at 9000 rpm. Spaggiari was absolutely consistent; he hit the same lines again and again. The session was over. No one had broken an engine, no one had dropped a bike.

Smart reported back. The first time out, the bike stepped around quite a bit on Dunlop TT100 street tires—once a racing tire was fitted on back, the machine settled down. There was still a wag in the works over fast bumps, but Paul reckoned that an external steering damper like the one on Spaggiari's machine would help a lot. How then did the machine handle? Not super-neat like the Triumph triple, but good and manageable. Paul liked the Ducati handling better than the works Kawasaki three he's been riding in the United States. Smart likes big bikes. He doesn't like to be cramped on a tiny machine which lurches and bounces and skips around corners. He doesn't enjoy a machine which develops a mind (and line) of its own around a 140 mph bend. The Ducati, stock frame and all, wasn't a headstrong bike.

Smart had a bit of trouble with the disc brakes which weren't set up quite right, but as for sheer stopping power, the bike had plenty. Fine, it stopped, but how did it go? That was hard to say. The engine was a willing revver—it would pick up at 6000 and pull to 9500 rpm so easily—it almost felt like there was another 500 revs on tap. Without a Triumph triple or a Kawasaki three on hand for back-to-back comparison, Smart couldn't really say: a Triumph triple felt faster—or maybe it just sounded faster, and the Italian twin was down ten, maybe fifteen horsepower on a good Kawasaki 750. But for Smart, any kind of comparison had to be tentative. Besides, horsepower comparisons at Modena seemed like pretty meaningless speculation; the real issue was lap times at Imola.

A big black cloud gobbled up the Modena sun, and it began to rain. Smart, his mind already on Imola, packed his leathers into the back of Vic Camp's car. "You know," he said in an almost off-hand way, as though he were talking to himself, "this might all work out very well." ◉

DUCATI 250 ENGINE

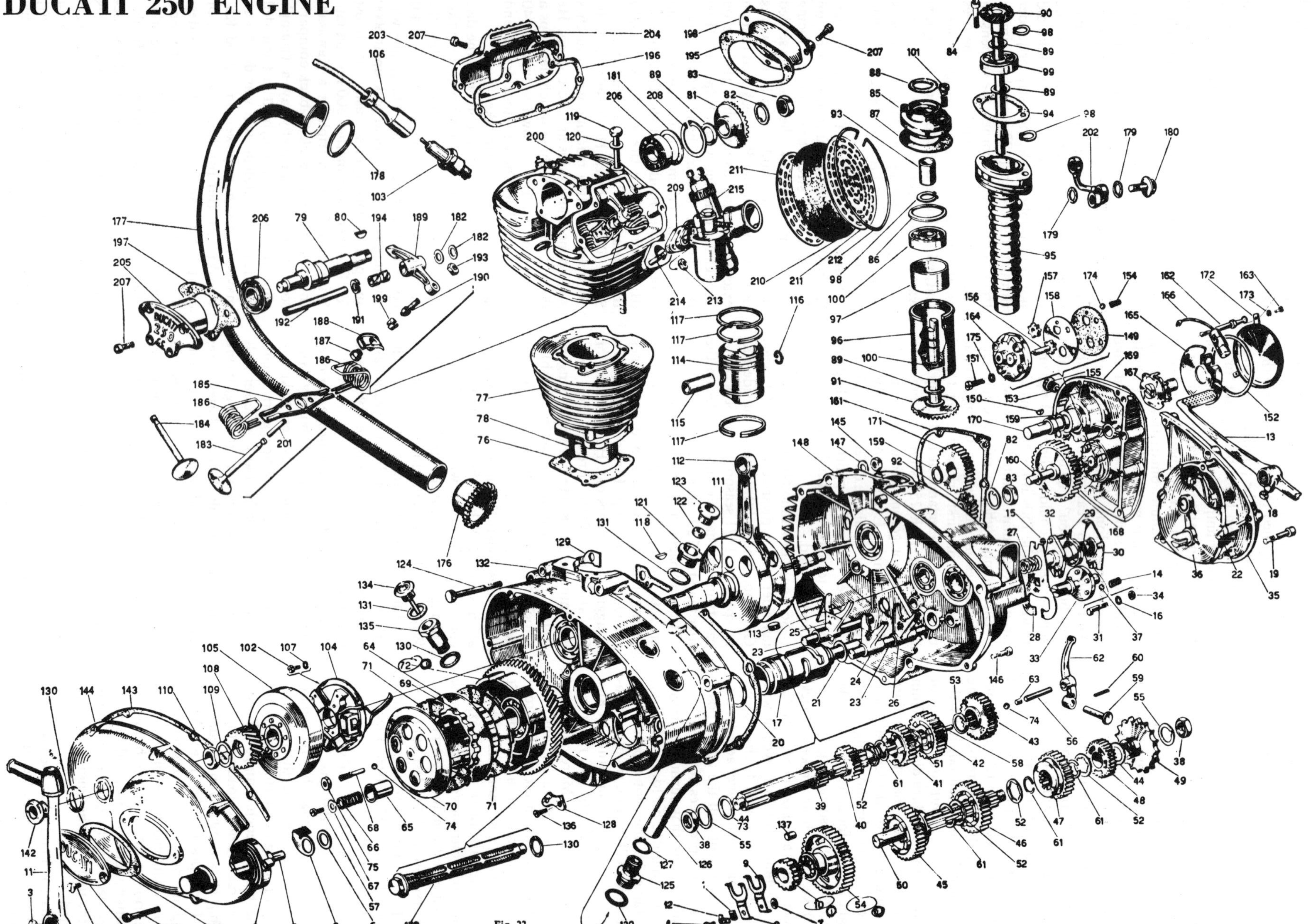

Fig. 22

The simplicity and strength of the design of the well-tried unit is apparent in this "exploded" drawing. The only point of vulnerability, says the writer, is the con-rod which must be changed once a season if 10,000 r.p.m. is regularly used

One lung breathes as well as two ...

RACING A 250 DUCATI

The writer on his Ducati at Thruxton in 1971

THE Vincent and the Velocette clans get on pretty well with each other. I don't suppose it is just because both machines are now without a manufacturer to their name, or because their respective designs have endured essentially unchanged from the dim and distant past: a better reason would be that both machines have a strong and distinctive character, so strong in fact that the whole machine is accepted, warts and all (and there are many of these), or not at all. This distinctive character is shared by one machine at least that is in current production—the Ducati.

It may surprise a number of the readers to learn that for the last year or so I have been racing my own 250 Ducati, and have grown to love it dearly. I waved goodbye to it recently to help raise the wind for a more modern racing machine, but not until it had borne me to the finish of many races with remarkable reliability.

It started life as a Mach I 250, but never made it to the road, as Vic Camp set it up as a Club racer for its first owner. The state of tune was mild indeed by the time that I purchased it, and embraced twin-plugs in the head, racing tyres and alloy rims, and a 30mm Amal concentric carburettor. The whole machine was in very good order, and had clearly not been used for a season or two. Just how old the tyres were was something I discovered only the hard way, late in the first season.

The standard Ducati front brake is only a single-leading-shoe affair, but is good enough as a start. And a start it was, for I had never raced a machine where the handling was the only thing that would allow me to get on terms with other bikes: the Honda 750—and its predecessors in my hands—had always had excellent acceleration. The Ducati was slower than most machines in Club races, and only by really riding properly could I expect to get anywhere. I therefore had to learn to do just that. The first problem was one of passing. When one is riding a fairly slow machine, with so-so acceleration, to pass someone requires rather more skill than needed to wind on the throttle on the 750 Honda and hold the backend slides . . . .

It is a fair tribute to Kevin Sheehan—from whom I had bought the Ducati—that I simply changed oil and plugs and took it to a BMCRC meeting to see how it went: everything seemed to be in accurate adjustment, and out I went. The contrast between 750 and 250 was so dramatic that the change from one down/four up on the left-hand side (Honda) to one up/four down on the *right*-hand side (Ducati) caused no difficulties on this occasion. I came away well pleased; I had been lapped in each race, it is true—but had finished in front of a lot of people and rated a place in the results each time. The lap speeds were pitifully slow; at 68 seconds one could hardly claim to be doing much racing.

By the next occasion that I raced at Brands Hatch I had learnt the simple lesson that a total loss ignition system has a finite—and limited—capacity: *ergo* a spare charged battery went into the toolbox. Sounds simple? Certainly . . . but somehow one just has to learn the hard way at times. On this occasion the Southern Sixty Seven Club were running the meeting, and the weather was wet and dry in spells. The first race was in drizzle on to a fairly dry track. The best time dropped to 67 seconds, though, and in the next race on a streaming wet track, drying out slowly as the rain stopped, the time dropped to 66 seconds. The handling of the Ducati was excellent, but the swinging arm seemed to be a little whippy: the brakes were beginning to seem efficient in their power . . . clearly the discipline of the low-powered machine was beginning to pay off a bit. At Thruxton next day the swinging arm let me down: having a really enjoyable dice with a Suzuki at about 1m 56s lap times, I looked down at Village Curve on the last lap to see a good half of the swinging arm coyly extended out of its retaining bushes. I stopped rapidly, ably assisted by a broken condenser lead which decided to let go at the same instant. Amazing the things that can go wrong. A swing arm retaining system was obviously essential. This was made up, and a CB92 brake fitted to make up for the distinct shortage of braking power. A new shortened skirt piston also went in, in readiness for the Bantam Club Enduro. This is a one-hour race, and therefore just long enough to be really called a race by my standards. This of course leaves me only the Isle of Man races and 24-hour Production marathons to call races, so I trust you will not press the point!

The first 10 laps were spent at a leisurely pace, to bed in the new piston, and from then on I bent my best efforts to a vain attempt to catch a very fast 125*c.c.* Yamaha. In the latter stages of the hour I found myself on very close terms with a Suzuki that had far better acceleration, and was a little faster. On any occasion when I got past or level that acceleration got me once more. The electronic timer on the Norwich straight gave a speed of exactly 100.0 through the trap for both the Suzuki and my Ducati, a very respectable speed for such a lightly tuned machine with such a heavy (13 stone) rider.

This kind of race is ideally suited to the Ducati, and the longer it went on the better it travelled. The final placing was a lowly eighth, but achieved with more enjoyment than any other race that I have ridden. It was after this meeting that I decided that the Manx Grand Prix was my ideal race: the steady and sustained pressure imposed by a long, long ride is really

RACING A 250 DUCATI

Clubs should cater for these strident little bangers

to me what racing has to offer. The last milligram of lateral gee pressure needed to get results in 10-lap short sprint races is not at all to my taste. The longer the race the more room for common sense and judgement, and the less the start matters. Just getting to the end of the race at racing speeds is a material achievement in races as long as the Manx GP. Indeed I hope to ride there this autumn. It is depressing to note that the best laps were as slow as 2m 6s at Snetterton on this ride, when Dave Forrester's desmo Production Ducati could consistently manage 2m 2s—let alone what the really fast machines could do . . . of the order of 20 seconds less again than that.

There is clearly a place for a framework of competition for the numerous Ducatis, Aermacchis and so on that are still to be seen in Club racing.

Winter saw a great number of modifications to the machine, none very expensive, but cumulatively effective. The little 30mm Amal was removed and replaced by a 32mm Gardner and an Amal Matchbox float chamber, the mild camshaft replaced by a new full race shaft from Mick Walker, and an alloy tank made up by Dennis Evans to replace the horrid fibreglass affair that had clung to the machine like a blood red barnacle for several years, new racing tyres went on, and the exhaust system was welded into one piece and re-routed under the engine to emerge snugly beside the sprocket on the opposite side to its original routing. This materially improved both ground clearance and accessibility and banished the cramped feel of the gear actuation pedal. The gear actuation was changed over to the left-hand side by routing a change bar through the swinging arm spindle on needle roller bearings. The six volt ignition was ditched and a 12-volt system installed with a Furakawa battery of substantial ampere hours and Siba coils. Tomaselli clip-ons and levers replaced the slightly flimsy and underdesigned Ducati units, and heavy duty cable was made to suit. The swinging arm was fitted with new bushes and re-reamed to give a really good fit. The front forks were overhauled and new staunchions fitted.

The net effect of all these minor alterations was to make the machine much tauter and responsive. The Gardner carburettor was strikingly effective in improving flexibility, in spite of the larger bore of the instrument. The advance-retard unit was welded up as a result and the overall flexibility and usefulness of the motor remained unimpaired.

The Triumph Owners Club High Speed Trials at Lydden Hill were an ideal venue for sorting out the carburation. Clutching a handful of needles for the Gardner and a set of sprockets for the circuit (on which I had not previously ridden), we went to Kent to find that the circuit had only just been resurfaced and the surface itself was still soft and covered with loose dust and small stones. It was very slippery and not a little tricky. On every outing both the carburettor and the gearing were altered, which did not help the riding a great deal, as the gear-change points altered on every lap in consequence. However, it was delightful to come up behind a big Norton or Triumph in a corner and see him edge away on initial acceleration before swiftly overhauling him as the Ducati got in its stride. That one-lunged acceleration is most deceptive. As a direct result of the bad surface the requirements for awards were cut down, and even with the lap penalty imposed for racing machines the Ducati managed to get several awards.

This was the last really enjoyable outing for some while, as coils broke down, rockers—subjugated to the higher stresses of the new cam—failed after many years of faithful service and finally snapped off. Ready once more, and handling beautifully with spot-on carburation and good braking . . . and then news of the early delivery of my new machine came through! I simply have not the time to keep two racing machines in the state of preparation that I demand from myself, and so the Ducati had to go. If it had stayed the temptation to race it would have been quite irresistable, and it soon went to (I hope!) a buyer who will use it as it was designed and developed to be used.

My 15 months of Ducati racing have—overall—proved to be reasonably cheap and much the same as Peter Gripton's (see recent issue); the final costs were higher as the machine was fitted with new parts and mods. up to the day before its sale, but allowing for that there was not a lot of difference.

The forgiving nature of the Ducati was underlined at an SSSRC meeting where it got its best placing (sixth). It was sliding and slipping all the way round Brands Hatch while the other machines did not seem to be finding any lack of grip at the same speeds. I then checked carefully on the tyres and found them to be several years old and far too hard for such antics. Nevertheless the precise handling of the Ducati kept me in full control—or at least effectively so—on that greasy track.

The basics of the Ducati engines are very simple. An exploded diagram is shown which illustrates clearly the simplicity and strength of the design. It cannot be seen in the picture, but the big end is a massive roller bearing affair fully in tune with the all-gear massiveness of the rest of the power plant. The primary drive is by bevel gears directly from the crank to the clutch housing on which the secondary gear is cut. On the timing side a direct gear on the crankshaft drives an idler well to a timing gear on which the points are positively located. A bevel drive gear is keyed to the crank and drives another bevel keyed to the camshaft drive. The cam itself is driven by a final pair of bevels, the final one keyed to the cam itself. This whole assembly has punch marks that are used for alignment, and the timing remains stable to better than a degree through repeated disassemblies and re-assemblies. The whole timing system for both valves and ignition is foolproof but for one small point. The ATD unit with the point cam can be fitted on 180 degrees out. It is only too frequently that one does this . . . but as it takes about a minute to correct it causes no difficulties.

The only point of real vulnerability is the con rod. Under racing conditions this must really be changed once a season if 10,000 r.p.m. is regularly used. As the peak power of the Ducati, as now set up for this season, seems to be at about 9,200 r.p.m. this has not really worried me (as I have a set of cases, a spare new con rod and a complete crank assembly ready to go in, that must be qualified a little).

My pet hatred on this motor is the vertical splitting of the cases. I like my motors to be oiltight, and this Ducati has certainly proved to be that: vertically split cases make this a lot harder to achieve . . . and this is an excuse that the Velocette owners might like to use from time to time. The key element in the fight against oil leaks was simple. Remove all gaskets, dress any mating surface if required AND USE A TORQUE WRENCH. This really did the trick, and when I got a socket-drive allen key this even stopped a weep from a rocker cover.

Lubrication is critically important on a racing machine, and I have found Shell Super H (a castor based oil) very good indeed. The motor strips clean, cam and follower wear was visibly lower than before, and what more can one say?

The Ducati as a road machine is a fussy beast that is always eager for an open road to clear its plug and get up on its high gearing: on the track it is manageable, tractable and forgiving, a very suitable machine for learning the trade. It is also pretty light. My Ducati weighed 241 lb filled with oil and with enough petrol for a 10-lap race.

The ordinary road machine is very much heavier and cannot approach the 220-230 lb dry weight of this racer. It is nevertheless a splendid machine with all the attributes of this one-lunged racer, and I shall be trying a 24-hour Ducati in full PR trim after it has completed in the TT. The contrast will be interesting, not least in the direct comparison of mid-60s and early 70s Ducati performance and handling.

I hope that clubs can be persuaded to cater for these strident and lusty little bangers in the future, and that these words might help to show why I think this would be a good thing, especially for novices in the racing sport. M.R.W.

THE REPACKAGING of the motorcycle for the mass market is of great benefit, I suppose, but sometimes has resulted in a decline of brand individuality.

There is a right way to do things, I know. Trouble is, when the masses are involved, manufacturers seem to use the same way even when it is different. Everybody, that is, except the Italians, who have found at least three right ways to do things, according to my latest count.

One of them is Ferrari, one of them is Fiat, and one of them is Ducati.

Ferrari, the animal which nobody can afford and few can tolerate. Fiat, the animal which anybody can afford and all can tolerate.

And Ducati, somewhere in between.

Trouble is, Ducati hasn't been noticed much by non-enthusiasts; for, after all, who bothers to notice the maker of nice little Singles with putt-putt displacements? Even a big banger like the 450 R/T fails to turn the head of the masses, though it sports such sophisticated four-stroke niceties as a desmodromic single overhead camshaft. Ducati is the only brand to regularly use desmodromic valve actuation to power the valve closed the same way you power it open.

The Ducati Diana 250 was the first production bike of note from Europe or Great Britain to have five speeds. It and its larger buddy, the 350, were screamers for their time—cafe racer's fare. Considered in the light of four-stroke Single technology, these subtly wonderful machines were one of the few—until Honda came along—that didn't dribble oil on your boots, garage floor or worse, the back tire. If treated with respect and a firm starting kick, they would run fast and far.

Ducati's sales in the United States don't amount to a hill of beans, compared to the Japanese. But, at last, Duck owners have a model carrying their brand name that will draw more than condescending (or vague, if at all) recognition, no matter how many it sells. That's because Ducati's new 750cc V-Twin is as animal as anything you can buy these days. It sounds tough, looks tough and runs as fast as it looks.

By the time you read this, more than a 100 of these machines will be in America. I was very fortunate to have an old friend call me up from Northern California to announce that he had just acquired a new toy.

When he told me what it was, I crammed a tape measure, bathroom scale, slide rule and camera into a sack and was on my way out of Orange County Airport with all those nine-to-fivers sipping dollar scotch out of paper cups as they contemplate the prestige of a 500-mile commute. I paid my buck, too, and settled down to wondering how weird life was that I should be commuting North with a canvas sack and a helmet bag....

My friend, Ron Paelutz, is kind of a "heavy dude," as they say these days. He's quite monied, but doesn't brag much about it. He paid the required 2000-plus dollars plus shipping for the Ducati because he wanted something solid, metallic and sporty. It had to handle, so while he was tempted to buy a Sportster and re-rig it as a lights-on sort of road racer, he felt that he would be closer to the desired combination by starting with the big Duke.

After sampling Ron's bike, I tend to agree. If you are a sporting (oh heavens, here come the letters about lawlessness) rider, there are some things that you are willing to overlook in favor of other things.

Like the gear ratios, for instance. The Ducati 750 is geared to do about 122 mph at its 7800-rpm red-line, in top (fifth) gear. This is not overly high gearing for a machine of this displacement, yet the bike will do nearly 50 in first gear. As

PREVIEW:
THE DUCATI 750

Come Take A Ride On A Formidable Cafe Racer! TEXT AND PHOTOS BY DAN HUNT

you can see, the gear spacing is quite close. So shifting the Ducati up is a steady, rhythmic succession of buttery smooth, small rpm drops. The spacing is beautiful, so that you can dial in a throttle setting in the right hand, then set your mental clock to shift in steady sequence after you pull away in that rather tall first gear.

The Feds haven't gotten to Ducati quite yet, for the transmission shifts in that traditional right-hand lever pattern—one up for first, and then down for each gear until you reach fifth. Neutral, as usual, is in between first and second.

The pre-ride briefing took place in Ron's home in the hills above Calistoga. It's a beautiful place from which to begin a ride, as it is nestled in the sparsely populated Napa Valley. On the west side of the valley, you can sample yourself rotten on free glasses from California's finest (Beaulieu, Charles Krug, Heitz, etc.) wine makers. On the opposite side, however, you can ride yourself dizzy on and off the Silverado Trail, a sinuous two-laner that seems to have been ignored by the tourists.

Ron gave me the usual cautions about taking it easy, as the machine had only 500 miles on the clock. What he meant was, keep it under 5500 rpm, which is about 85.

Walking around the bike, I could see much that is typically

DUCATI 750

An international panel, with Italian labels and English instruments.

Ducati. Like the finned, deep, 10 pt. capacity oil sump. This seems large, but perhaps Ducati feels that it must circulate that amount of oil to counteract any possible inadequacies in air-cooling imposed by the rearward position of the No. 2 cylinder barrel.

This engine is a 90-degree V-Twin and, as the front pot is nearly horizontal, the rear cylinder receives a fair amount of air circulation compared to the H-D V-Twin, which has only 45 degrees between fairly upright barrels. Italy's other 90-degree Twin is the Moto Guzzi, and it avoids the cooling problem entirely by turning the engine so that both barrels are in the airstream and using shaft drive.

Ducati's philosophy is the more sporting of the two, for it allows use of slightly more efficient rear chain drive and presents a very narrow engine profile for the rider. The center of gravity is quite low, because of the forward cant of the front cylinder barrel. You can argue all day about the fact that you'll never drag the pots on a shaftie like the Guzzi or BMW, or the cases on an in-line transverse Four, but you'll never convince me that the guy with the narrowest and lowest engine doesn't have at least a psychological advantage in the corners.

There are certain trade-offs in Ducati's choice, however. The fouling and shielding of a clean blast of air to a V-Twin's rear cylinder may be counteracted by superlative oil capacity and circulation which Ducati obviously has. But Ducati, having chosen 750cc and 90-degree cylinder disposition, must cope with a rather long wheelbase—60.2 in.

You can make anything handle well if you work at it, but that 60.2 in. will tend to slow the machine's steering responsiveness. The Norton Commando 750, another individualized vertical Twin answer to the problem of going fast, stopping and turning corners in neat fashion, has only a 57-in. wheelbase, and even Harley's 1000cc Sportster manages to hold the wheelbase to 58.5 in., probably because of its smaller cylinder angle. There is, after all, only so much room, per a given rake and wheel size, between front axle and swinging arm. The Ducati uses it all.

The V-Twin concept, of course, has a great history to it. As far back as there were big motorcycles, there were big V-Twins. Harley, Indian, Big X, Vincent HRD, etc. Designers glommed on to the V-Twin for the ergonomic reasons we have already outlined and then for more subtle reasons. When they tried to improve on the V and build Fours (Ace, Henderson, etc.), they found their technology wasn't good enough to make the Fours acceptable, reliable or even profitably producible.

So the V held on, reviled by some for its grossness and staggered firing pattern, and yet loved by many more for its relative simplicity and even its raw look of bigness. The technical reason for the V is its theoretical superiority of mechanical balance over the parallel Twin in which the pistons run up and down in unison. This sort of parallel Twin is hardly better than a big shaky Single.

Yet the V-Twin, especially in the 90-degree configuration, offers the promise of total elimination of primary imbalance. There is left a secondary component that averages out a nearly horizontal reciprocating thrust in line with the machine, but hopefully the elimination of primary imbalance will suppress those strong resonating impulses which course through the motorcycle frame and mysteriously crack fender mounts and the like.

If all the theory in the world won't sway you that you can build a relatively vibration-free V-Twin, you're right. When I fired up Ron's Duke, after an inexperienced tickle and four nerve-racking kicks, it settled down to that gorgeous staggered

The 750 engine is a single overhead cam 90-degree V-Twin. This model does not employ the "desmo" head.

Ducati's front disc brake employs a hydraulically actuated caliper. Response is excellent.

Removal of seat reveals tool tray, tools, battery box, tank fasteners. It's not exactly quick detach, or replace.

A bullet-proof rear axle adjuster, and a worm's eye view of Ducati's rough-hewn metal work.

lope like every other V-Twin I've ever ridden. That is, the bars tremble and your butt jiggles, letting you know that there is one big, bad machine churning away underneath you. A machine.

While the Ducati was warming, I was briefed on the controls. The white light on the non-glare, rough finished black instrument nacelle tells you the ignition key—in a clumsy spot under your left thigh—is on. The green light on the right says you're parking lights are on, and the left hand red one indicates high beam is operating. They haven't bothered to translate the light functions for the American consumer and so these lights are labeled with abbreviated spaghetti, "Gen., Abb., Luci," just like when you go to Modena to pick up one of Enzo's creations. The light switch is also on this nacelle, which carries a brace of Smiths instruments; these are in English and mph. In general, the switches are of mediocre quality.

This part of the instrument package is neat, but the issue is confused by the startling use of chromed, heavy-gauge spring wire to mount the headlight. A proper guess is that use of the wire makes it very easy to mount short racing bars on the fork tubes where they're supposed to be, without disturbing the light brackets.

The speedometer and tachometer are individually lit and this street racer even has turn signal pods. The control cables and the wiring for all these lighting conveniences are not disguised well, and could stand some studious taping and rerouting, or, in the case of the turn signals, simple removal. The indicators might as well be removed, as no one, including Ducati, has yet figured out a turn signal that is convenient to operate and self-explanatory in operation. The car makers use a stalk, knowing that a proper driver makes frequent use of the indicator when in traffic. Thumb switches on bikes are awkward. Perhaps a twist-operated spring-return two-way switch in the left-hand grip is the answer.

The seat of the Ducati 750 is typically Italian and thus is stiff, even if it is quite wide. Its saving grace is that its width is matched with the slightly scalloped fuel tank, so that, when you take off, your knees come together naturally to a firm comfortable grip on the tank. Underneath, if you have the patience or motivation for the unscrewing ritual, is a large tool tray with a shoddy looking assortment of stamped tools. Quick replacement of the seat is directly contingent on how sly you are in fitting it to two alignment rods on the frame. Neither Ron nor I were very sly, it seems.

Fiberglass is used for the fuel tank and side panels, and the finish is attractive but poorly executed. There are small pinholes in the gel coat. This is nothing new, for the fiberglass tank on the Ducati 450 R/T we tested had the same level of finish. One improvement is that, unlike the 450, the 750's gasoline petcocks are easily accessible.

I pulled off gingerly and idled down the hill. Ron fell in behind on this fantastic old Fifties Triumph Twin with rigid frame that he'd bought for $50. It was a spare bike to keep at his vacation home, and he'd been running it for two years straight with little more than an occasional oil change.

The Ducati was surprisingly easy-going at crawling speeds. At 32 in., the seat isn't very high, so the bike is no chore to handle at a stoplight. And, as soon as it's moving at all, it self-stabilizes in a gyroscopic feel reminiscent of a Harley 74 or BMW. One reason for this pleasing behavior is that low center of gravity afforded by the oil sump and nearly horizontal front cylinder. In a way, it's the same thing that made the H-D Aermacchi Sprint such an incredible handler.

In spite of a set of formidable valve timing specifications, Ducati's single cammer pulls off from a stop neatly with no erking and runs well at slow speeds. The abovementioned timing includes a great deal of overlap. The figures are: inlet opens 65-deg. btc, closes 84-deg. abc, with a specified tappet clearance of 0.0020 to 0.0039 in.; exhaust opens 80-deg. bbc and closes 58-deg. atc, with tappet clearances of 0.0039 to 0.0059 in. This yields an overlap figure of 123 degrees, equivalent to a fairly ratty Norton Manx road racer. Most big bore road bikes for sporting use have more moderate overlap—usually in the 80s or low 90s—assuming that the machines will not be used at high rpm most of the time.

The mitigating factors in this case are the use of moderate 8.5:1 compression ratio and conservative 30mm carburetor choke size. The reason that Ducati uses wild cam timing has to do with that "other" model—the production racer designed to drub MVs at Imola, turning an impressive 9000 rpm all the while. So the machine runs in reasonably docile fashion, albeit yielding unimpressive fuel consumption between 30 and 35 mpg. Overlap wastes a lot of gasoline.

Once we reached the Silverado trail and pointed ourselves toward Napa, I got an indication of what a fine roadster this is. It's what the Italians call "gambelunghe," or long legs. After a brief period of marked vibration at about 4200 rpm, it settles down and wants to go fast. Without a lot of busy work. Singles have that quality, as do BMWs and properly geared Nortons and Harleys. Now you can add Ducati to the list.

The bike stops and soaks up road undulations well. The classic test in the Napa Valley is to leave the Silverado trail at the Lake Berryessa road and zoom up a narrow canyon road straight out of the mountain climb at the Isle of Man. The road is smooth but swervy until you pass a reservoir and switch right into Lobo Canyon. The first section tells you about brakes, because you have to actuate them while the bike is heeled over. The second section gives you more of the same, but tighter, and also tries to smash your forks up through your teeth.

The Ducati feels properly stiff without being rock hard. In spite of the long wheelbase, the bike showed little tendency to plow in the 45 to 60 stuff. It is, after all, relatively light at less than 440 lb. The front disc brake is quite strong, but is easy to control delicately. Lever travel diminishes greatly once the pucks are gripping the disc, but it is still possible to feel how much more or less pressure is needed. On rough pavement or during heavy braking, the Ducati doesn't heave as you might expect some machines of this size category to do. My sole complaint has to do with the narrowness of the tires, particularly the back one. Granted they track well, but with the horsepower that this 750 develops, Ducati would do well to go to one of the larger super-compound tires developed for the Japanese big bores. As for the front tire, a larger width based on a road racing design may be necessary to eliminate possible plowing arising when bike is pushed really hard.

The handlebars are overly wide for spirited riding, and not too appropriate to the bike's natural proportion. The megaphone-styled but relatively quiet silencers are raked nicely upward to provide excellent turning clearance. The first place you'll hit when turning left is the center stand.

Due to the newness of the machine and the lack of a dragstrip nearby, I can't give you concrete acceleration figures. But a glance at the computed speeds in the various gear ratios will be revealing. I think the bike is geared quite honestly.

The gearing figures will also tell you that this machine is not a drag bike in delivered form. If you want that, you'll be better off with a Kawasaki 750 or even the Norton 750. My computations tell me that the Ducati will do a quarter mile in no better than the middle or high 13s, and will run through the traps at about 95 to 97 mph. Regearing would help the e.t. considerably, but not change the trap speed much.

The man who deserves to own a Ducati 750 will not be speculating about this sort of straight-line statistic, anyway.

For a Ducati owner, the shortest distance between two points is a curved line. ◙

DUCATI 750

SPECIFICATIONS

List price	$1995
Suspension, front	telescopic fork
Suspension, rear	swinging arm
Tire, front	3.25-19
Tire, rear	3.50-18
Brake, front, diameter x width, in.	11.0 x 1.44
Brake, rear, diameter x width, in.	7.87 x 1.66
Total brake swept area, sq. in.	116.1
Brake loading, lb./sq. in.	5.93 (150-lb. load)
Engine, type	four sohc V-Twin
Bore x stroke, in., mm	3.15 x 2.93, 80 x 74.4
Piston displacement, cu. in., cc	45.7, 748cc
Compression ratio	8.5:1
Estimated bhp @ rpm	60 @ 7800
Claimed torque @ rpm, lb.-ft.,	N.A.
Carburetion	(2) 30mm Amal 930
Ignition	battery and coil
Oil system	wet sump
Oil capacity, pt.	10
Fuel capacity, U.S. gal.	4.5
Recommended fuel	premium
Starting system	kick, folding crank
Lighting system	12V battery, alternator
Air filtration	paper element
Clutch	multi-disc wet
Primary drive	gear
Final drive	single-row chain
Gear ratios, overall:1	
5th	4.80
4th	5.52
3rd	6.63
2nd	8.62
1st	12.33
Wheelbase, in.	60.2
Seat height, in.	32.0
Seat width, in.	10.4
Handlebar width, in.	32.2
Footpeg height, in.	11.7
Ground clearance, in.	6.9
Curb weight (w/half-tank fuel), lb.	438
Computed top speed in gears (@ 7800 rpm), mph:	
5th	122
4th	106
3rd	88
2nd	68
1st	48
Mph/1000 rpm, top gear	15.6
Engine revolutions/mile, top gear	3843
Piston speed (@ 7800 rpm), ft./min.	3808
Lb./hp	9.8 (with 150-lb. rider)
Fuel consumption, mpg	30-35

CYCLE ROAD TEST

DUCATI 750 V-TWIN

● When you see the Ducati 750, you suspect. When you ride the bike, your suspicion grows stronger. And after you've spent mile after mile, corner after corner in the saddle, you know it. You know that this motorcycle did not originate in detailed market research, that no armchair committee of twenty corporate honchos dabbled in the design, that the power of a mainstream idea was not bled off into a hundred different tributaries. When the disc brake squeezes the bike down from three-figure speeds, when the bike connects your nerve endings to the tire patches, when the rightside peg nicks down in an 80-mph sweeper and the bike never bobbles, when the 750 leaps forward from 3000 rpm in fourth cog—then you know. You know that a motorcyclist designed this machine, and he got it right. And motorcyclists built this bike, and they kept it right. Above all else, the Ducati 750 is a motorcycle. It is not an appliance. The people who created this 750 probably could not develop a good American refrigerator or food mixer or steam iron. And about such things, thankfully, they do not care.

Ducati wanted a 750, and they wanted a good one fast. Ducati had to be dead certain—before they began the project—that the end-product would work, and work well. There was no margin for gross errors, or even little niggling ones. There wasn't time and money at Ducati to do long-term development of a multi-cylinder design, and besides Ducati just wasn't convinced that transverse in-line multis were necessarily the ultimate answer for street motorcycles.

The 90-degree V-twin is basically two 350 Ducati singles lashed together. That bald description is a lit-

PHOTOGRAPHY: DOUGLAS MESNEY

(Left) Lockheed caliper bites down on a cast-iron 11-inch disc; brake is powerful and predictable. Disc rusts on water contact. (Top Center) Control switches are unfortunate: too many bits-and-pieces and too hard to use. And the plumber's nightmare at the end of the brake reservoir should be sheathed in rubber. (Top Right) Instrument panel is handsome and well-angled, but the pilot lights are much too bright at night. (Center Below) Sls Grimeca rear brake works smoothly; mufflers are beautifully shaped. (Right Below) Right fork leg has boss for a brake caliper.

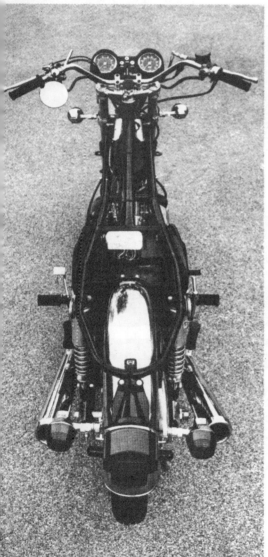

The frame is narrow and rigid; tube running from the top of the steering head neck hooks up with two cross-bridges which span main parallel frame tubes.

tle too simple; more precisely, the 750 engine is closely patterned after the 350 single, and both engines share the same engineering traditions and working technology. The bore and stroke of each pot in 750 is 80mm x 74.4mm, dimensions which approximate those of the 350 single (76mm x 75mm). The twin's cylinderheads follow the time-honored fashion of the Bologna firm: single overhead camshafts driven by helical-cut bevel gears and towershafts. The camshafts, rolling on ball bearings, operate rocker arms which have no screw adjusters for dialing in valve lash. Instead, winkler caps perched atop valve stems must be changed to adjust clearances (as in the old 250 Mark 3s). Mechanics who set valves by the minute-hand of the clock hate the cap system, but it is positive. There's no extra weight bobbing up and down on the end of the rockers, and there are no adjusters anywhere to come out of whack. Although current Ducati lungers continue with dated hairpin valve springs, the 750 twin, in keeping with more contemporary spring technology, uses coil springs.

The three-ring pistons gulp down fuel/air mix through two 30mm 930 (Spanish) Amal carburetors. The pistons don't pressure the charge too much by modern-day standards; compression is a fairly modest 8.5:1.

Like all Ducati engines, the twin is an all-aluminum alloy affair, and the main cases split vertically. The front horizontal cylinder is a bit offset to the left from the vertical cylinder, and downstairs on the crankpin the forward pot's connecting rod rides to the left of the rear cylinder rod. Both of the one-piece connecting rods turn on caged rollers at the big-ends, while plain bushings do the job at the wrist pins. Ducati elected to use a pressed-together crank assembly with one-piece rods running side by side because they were positive such an arrangement would create no past development problems. Ducati had no experience with male and female con rod setups or with bolted-up flywheel assemblies—the type which Harley-Davidson employs in their production V-twins.

The 750 Ducati's lower end assembly with its truncated flywheels rides on huge ball bearings. Outboard of the left main bearing and outside flywheel, a helical-cut primary gear passes the power to a wet clutch and thence to a five-speed transmission. On the right side of the engine, a set of helical-cut bevel gears, driving off the crankshaft, turns a shaft which spins an idler gear which in turn moves the lower drive gears for the camshafts. The shaft operating the idler gear continues upward and drives the contact breaker points. Tucked into the right side outer case is the 150-watt alternator and a vane-gear oil pump. The pump is driven by an idler gear which meshes with a spur gear on the right side of the crankshaft. Like small Ducati singles, the 750 is a wet-sump engine—the sump holds

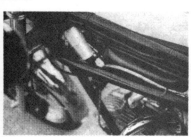

(Top Left) Rear brake light switch hangs out in the open; note the gussets at frame member junctures. (Left Below) Removing tank exposes ignition coils and air cleaner box for the forward cylinder.
(Top Right) Engine looks busy on the right side. Mounting bosses are at the end of the front downtubes—this method is a strong, precise and expensive way to frame the engine mounting bolts. (Right Center) Left side of the engine is cleaner, neater. The ignition key is at the top leading edge of the side cover which has frilly Flash Gordon air scoops. (Below Right) Condensers shove into the exquisite casting which surrounds the contact breaker points.

headlight is insulated from vibrations by rubber grommets. The entire headlight/directional light assembly is mounted on chromed spring steel wires giving the front end a lean, light look—besides saving bulb filaments.

One of the nice touches on the Ducati is the chain adjusters which operate inside the swingarm tubes, making for precise and rigid adjustments.

4.5 quarts—so there are no hoses, clamps, fittings, or outside oiltanks. And no leaks.

The alloy engine, with dozens of carefully matched gears whirring away inside, is an expensive thing to build. One might expect that Ducati, having spent a bundle on the engine, would have snipped a couple of corners and slipped the bucks-up engine into a money-saving frame. That didn't happen. The frame is built of seamless, chrome molybdenum steel tubing. Double downtubes drop from the steering neck and bolt to the front of the engine. Two frame members grip the engine from the rear; these rear tubes also provide the mounting bosses for the tubular swingarm. Just above the swingarm mount, the rear tubes run upward to meet two main horizontal tubes running back from the lower part of the steering neck. A long single tube connects the top of the steering neck to two crossbraces which bridge the horizontal frame tubes; one bridge crosses just above the vertical cylinderhead, the second bridge is several inches aft of the first one. The main horizontal frame tubes splay out (just under the riders portion of the saddle), travel back to locate the upper shock mounts and hook up at the rear of the bike. Support struts for the rear section of the frame join the upper shock mounts to the rear cradle tubes at a point just above the swingarm mounting bosses. The swingarm is noteworthy because the rear chain adjusters operate inside the tube itself—which makes for a very rigid and precise way of moving the rear wheel.

An 8-inch, single-leading-shoe Grimeca rear brake with a built-in cush-hub rides in the swingarm. Three-position Marzocchi rear shocks control rear wheel action, while upfront special built-for-Ducati Marzocchi front forks deal with bumps and ripples. The front wheel hub is a beautiful piece of casting, but the real attraction is the 11-inch front disc brake. The familiar Lockheed caliper unit bolts on the left fork leg; the pucks bite on a cast-iron disc. The right fork leg carries mounting bosses for a second caliper unit, so that a double-disc setup (a la Imola roadracers) is possible.

The Ducati 750 looks different. A snap impression would label the machine as ungainly, something akin to an overgrown dragonfly: beautiful in its component parts, but unattractive as a single piece. That sort of judgment misses the point. The big Italian twin is a basic example of form following function. It looks the way it looks because that's the way the machine in its central layout and main dimensions was engineered. Any styling that was done—or can be done in the future—with this motorcycle is all after-the-main-engineering-facts.

For example, take the 90-degree engine which is buckled longitudinally into the frame. Combined with a steering head angle of 29 degrees, the layout makes for a long (60-inch wheelbase) motorcycle, and a lot of empty spaces. A vertical twin would have been neater in appearance, but that didn't matter. Ducati wanted an inherently smooth-running engine which could be built within their limits of time and money. A 90-degree V-twin has perfect primary balance, and its rocking couple occurs at twice the crankshaft engine speed—all of which means that there's no vibration worth noticing. As a bonus, the twin-cylinder 750 engine is only a couple of inches wider than a 350 single. To avoid quaking and shaking at low revs, and to retain a simple final chain drive, the engine went into the running gear with its crankshaft rotating in the same direction as the wheels. So what you get is an incredibly smooth large-displacement twin, a motorcycle which needs no rubber-biscuit engine-insulating system, no tall vibration-damping gear ratio, no sponge-sprung handlebars or fatso foot rubbers, and no safety-wire and Locktite to stop nut-and-bolt absenteeism. The object was a smooth-running big twin. That was the result. The Ducati 750 looks the way it does because the central question was "how will it work," not "what will it look like." For all this expense, sophistication and form-following-function business to mean anything at all, the Ducati 750 has to work well—even brilliantly. The machine does.

The first thing you notice when you roll the Ducati off its centerstand is that it weighs a lot less than it looks. At 446 pounds with a full tank of gas, the Italian twin is the second lightest 750-performance bike around. (Norton's Commando is the trimmest muscle machine at 435 pounds). After you turn on the gas-taps, click on the ignition switch, shove the carburetor air-bleed lever forward, you're in for your next surprise. The kickstarter cranks the engine over with ridiculously low leg effort. You could almost start the engine—first or second kick everytime—by spitting on the kickstart lever. The mechanism, however, could use a slightly stronger return spring.

With the carb air-lever returned to its normal position, the idle settles down to a steady rumpty-rumpty-rump 800-rpm beat. The Smith rev counter and matching speedometer report all the vital numbers; the instruments, sunk in rubber, have little hoods over their tops, and this treatment reminded more than one staff member of frenched headlights on cars of the 1950s. The dials share the pod with a light switch and three pilot lights: a white generator light (marked *GEN*) for the alternator system; a red eye (*ABB*) for the high beam lights; and a green light (*LUCI*) which signals that the lighting circuit is operating. These little lights are annoyingly bright for night-time country riding—and they should be dimmed down.

Even though the clutch has a nylon lining twixt the cable and sheath, the clutch lever draws hard. Those enthusiasts accustomed to British machines will find the clutch draw requires "normal" effort. But those riders who have grown up on Japanese clutches will immediately develop sore wrist and forearm muscles from the Ducati clutch.

The twin produces a visceral, throbbing exhaust note below 4000 rpm—which turns into a heart-stopping warble at 5000 rpm and up. Whether or not these sounds sneak inside the legal dB(A) limits, we could not ascertain at test time because *Cycle*'s East Coast sound meter had gone deaf. The exhaust-note music, and the fact that the bike doesn't shake at idle or upstairs in the rev range, might keep lots of Ducati riders in town, doing a grand tour of all the hot dog and root beer stands. Of course the heavy clutch doesn't exactly encourage this sort of round-town rambling, though the engine is torquey enough so that you can be sloppy and even forgetful of gear changes. Even so, in-traffic steering is sluggish and heavy, because the frame geometry is calibrated for fast traveling on open and winding roads. Once you discover *that*, and once you realize how good the Ducati 750 is for fast road riding, you may never want to waste the bike in town again.

In the past three years, *Cycle* Magazine has found two motorcycles which have been about equal in terms of the enjoyment which they gave the testers when riding hard on mountain roads and straight pikes. Those two machines are the Honda 750 Four and the Benelli 650 Tornado. By a small margin the Ducati 750 is more enjoyable to ride really hard than the big Honda and vertical-twin Benelli. Although the Ducati's V-twin feels fairly clumsy at low speeds, at high speed the machine is stable and predictable; it never thinks about doing any tricks in fast bumpy corners. It doesn't hop or jump or shake. Nor do sidewinds bobble it. And we didn't even screw the steering damper down. On the left side, the centerstand grounds before the left footpeg, and you have to be motoring right along to touch down the stand. On the right side, you can dive into corners and just keep going and going until you feel the fold-up footpeg dragging away on the pavement.

Tire pressures were about 30 psi in both Metzeler tires. The tire compound seems

DUCATI 750

Price, suggested retail	East Coast, POE $1995.00
Tire, front	3.25 in. x 19 in.
rear	3.50 in. x 18 in.
Brake, front	11.02 in. x 1.5 in. x 2 in.
rear	7.87 in x 1.15 in
Brake swept area	132.2 sq. in.
Specific brake loading	4.7 lb/sq. in., at test weight
Engine type	90-degree V-twin with two single overhead camshafts
Bore and stroke	3.120 in. x 2.902 in., 80mm x 74.4mm
Piston displacement	.45 cu. in., 748cc
Compression ratio	8.5:1
Carburetion	2 (#); 30mm; Amal 930
Air filtration	Pleater paper element
Ignition	Battery and coil
Bhp @ rpm	N.A. (hp) @ N.A. rpm
Mph/1000 rpm, top gear	15.3
Fuel capacity	4.5 gal.
Oil capacity	9 pints
Lighting	12v, 150 watts
Battery	12v, 12ah
Gear ratios, overall	(1) 12.32 (2) 8.60 (3) 6.63 (4) 5.51 (5) 4.89
Wheelbase	60 in.
Seat height	30.5 in., with rider
Ground clearance	6 in., with rider
Curb weight	446 lbs., with full tank of gas
Test weight	626 lbs., with rider
Instruments	Speedometer, tachometer, trip odometer
0-60 mph	6.0 seconds
Standing start ¼ mile	13.93 seconds 95.00 mph

DUCATI 750

fairly hard, but they do grip well (at least in the dry), and in this respect, the Metzeler tires remind us of the Michelin Supersport street tires. Ducati 750s which we've seen have had either Michelin, Metzeler, or Dunlop TT100 tires.

The Marzocchi front forks and rear shocks are excellent. It's clear where Ducati has spent their money building this motorcycle. There is, after all, no deep secret anymore to build good suspension components; assuming that someone has determined the proper spring-and-damper rates for a given machine, the real trick is getting accurate, consistent machining and careful assembly. And that costs money.

Another place where patient assembly and close tolerances pay you back in decent handling is in the setup of the steering head bearings. In our test bike, everything was perfect. There was no slop whatever in the system, and no binding either. Once upon a time, especially in British circles, there was much talk about the undesirability of frames which employed engines as an integral stressed member; according to the oracles, such jerry-rigging got the tubes 'agalloping and the steering with it—no matter how "right" the geometry figures were. Full cradle duplex frames, Manx Norton style, were supposed to be the only way to get a rigid structure and proper steerer. This eternal tenet has since passed to dust, and the Ducati 750 frame suggests why. The twin's frame, which is an abomination according to the old gospel, steers very well indeed. And it should. The frame succeeds in holding in sufficiently precise alignment those parts (steering neck, forks, swingarm pivot, and axles) which describe the frame geometry.

The engine itself makes going quickly easy. It produces a lot of torque low down in the rev range—wicking it on at 3000 rpm in fourth gear produces a driving surge forward. Perhaps the most outstanding single characteristic of the Ducati 750 is the abundance of engine torque. The Ducati is more fun to ride, for example, than the limited production Moto Guzzi V7 Sport thanks to the Bologna twin's wide power band. Ultimately, an experienced rider can travel more quickly from point to point on the Guzzi V7 Sport than the Ducati twin. But the Guzzi makes you work at it.

A few runs at the drag strip revealed that the clutch on our test bike was a bit out of adjustment (too tight), and for that reason, we got some clutch slippage. The bike toured through with 13.9s at 95 mph. The engine felt as if it were all done and out of breath at 7200 rpm; pushing the tach round further did not improve times. Ducati can make their new V-twin go a lot faster and quicker by building in more horsepower, but paying for that sort of seldom-used performance by trading away the present engine's tractability and docility seems stupid. The real joy with this motorcycle is pitching it into a corner, getting on the twistgrip in third or fourth at 3500 rpm, dialing in the power and running the engine up to a 7000-rpm warble. The smoothness of the engine allows the rider to exploit the bike's power fully. Our test bike had some initial vibration right at 4000 rpm, but this disappeared upward into higher rpm figures as mileage accumulated, and eventually the momentary quivering vanished altogether.

Over long stretches the motorcycle stays pretty comfortable. The seat is solid; shape and width are fine. Out test riders, settling into the saddle pocket, found it comfortable. The pocket lowers the seat height to an acceptable 31 inches. (Remember that the bike physically is big and tall.) The seat padding is just a little too stiff. The space relationships between bars, pegs, and seat are pretty good—good enough to make five straight hours in the saddle an acceptable proposition.

Neither our 6'1" or 6'0 test riders particularly cared for the handlebars which are three or four inches too wide. The brake lever pulls easily, and it feeds back messages to your hand from the tire. The Lockheed disc is powerful enough to squeal the front tire on demand, but it's predictable in the same way that a Honda disc is predictable. The cast-iron disc doesn't squeak, unlike stainless-steel discs which Japanese bikes carry. The Ducati disc, however, rusts immediately after getting wet, and stays rusty.

The rear brake pedal is a little awkward to use; you have to consciously lift your foot up and around to get on top of it. The rear brake action is smooth and steady and the shoes showed no signs of wanting to grab. The gearshift lever is positioned better than the brake pedal. Your boot doesn't have to come off the peg to stab it down, and pulling the lever up is simple—the lever rides right at the top of your boot. The actual shifting is traditionally Italian; the throw is a bit longer than most Japanese bikes, but it changes gears so positively with a *clucka-click*.

The handlebars contain most of the really twinky pieces of equipment found on the new Ducati. The fittings which join the disc brake master cylinder with the hydraulic line look like an oil-well Christmas tree. The switches—dimmer and horn on the left side, and direction signals and headlight flasher on the right—are 1950-vintage Italian. And there is no positive kill switch on the handlebar. In order to incorporate the vintage switches with the new Lockheed handlebar lever/reservoir unit, more bits and pieces became necessary. Worse, none of the switch controls are easy to use; they are just too far away from the thumbs. Clearly, the Ducati needs a completely new set of rationalized, integrated handlebar controls, and we see no reason why this can't be done. After all, the Italian CEV company managed to design and produce a new taillamp which meets American standards. Surely *someone* in Italy should be able to produce complete handlebar control units. Bits-and-pieces handlebar controls just aren't acceptable on motorcycles in the price and quality range of the Ducati 750.

Neither is the poor quality toolkit. Just about the same toolkit came standard on the first Ducati singles imported into the United States about 14 years ago, and an upgrade change is years overdue. We have two further complaints about component quality: some of the fiberglass works (tank and sidepanels) showed pinholes, and in general, the fiberglass work is not in the Rickman class. Much more aggravating is the fact that the decal striping on the tank started to peel away when the motorcycle was washed with a car-wash jet-gun. And a half-dozen allenhead socket-screws, odd little springs and cylinderhead bolts (as well as the cast-iron disc) rusted after their first contact with water. Clearly, Ducati should take immediate steps to rectify these shortcomings.

These items should be upgraded because other components of roughly the same importance are first-rate pieces. For example, the fenders are stainless steel. Borrani alloy rims are standard. The 12-volt electrical system employs a Japanese Yuasa battery. The headlight and directional signals are mounted on spring-wire brackets, and the taillight and headlamp are rubber-shock-mounted. The instruments swim in rubber. Even the horn gives off a healthy croak, and that's far different from the old Italian horns of yesteryear, which in an emergency could only manage a feeble gurgle. When you start adding nice touches, you really shouldn't quit before it's all perfect. A motorcycle which carries Borrani alloy rims as standard equipment just should not have a single rusty allenhead socket-screw on it.

If we had to make a choice between good suspension components and nifty, slick cad-plated tools, then we would go for the suspension every time. But that is not really the choice. The choice is whether Ducati—and in the end the consumer—will pay the extra loot to upgrade the items in question. We think the Ducati clientele, at least in America, will. People who buy machines like the Ducati 750 aren't nickel-and-dime misers. To high-buck buyers quality counts. That's why they pay premium prices for premium mechanical things in the first place. And that's what they should get.

This year Ducati 750s will be in relatively short supply. So if you buy a motorcycle for the way it looks, or for its candy-apple paint scheme, or even for what it says about your manhood, then pass the Ducati 750 by and get something else. Let some down-to-the-marrow motorcycle enthusiast buy the Ducati 750, some guy who wants the big V-twin because he knows—and appreciates—that as a mechanical thing, as a piece of machinery, as a *motorcycle*, the Ducati 750 really works. ⦿

It's almost a tradition. When the Italians build a touring version of any motorcycle, they later offer a sporting model of the same basic machine. Thus, when Ducati introduced their new V-twin, the bike was outfitted in grand touring garb. But there was reason to believe that some sort of tuned version would subsequently materialize.

Those who followed the development of the Bologna bike knew that an early prototype V-twin at the factory carried 35mm carburetors and would spin to 9500 rpm; these were considerably bigger figures than the ones surrounding the final GT-750 production version which mounts 30mm carbs and delivers maximum horsepower at 7200 rpm. The factory had the goods—the only question was when they would spring (and sell) the tuned-up counterpart. The answer was late '72.

The Ducati 750 Sport is a factory-made café racer; it holds the same relationship to the standard touring version as the Norton Production Racer bears to the cooking Commando. Most of the Sport's major differences are things which one can see: fiberglass tank, seat, fenders and half-fairing; clip-ons, remote shift/brake lever assemblies, electric tachometer, black engine side covers, rubber headlamp stanchions and a slightly different taillight.

Mechanically, the changes are pretty minimal. The outstanding modification is a pair of 32mm Dell'Orto pumper carburetors, which open to the atmosphere through intake trumpets and wire-mesh screens. The screening will strain out anything larger than medium chunks of gravel. The absence of any air cleaners and their accompanying plumbing will automatically improve performance, as *Cycle* discovered in its Superbike Comparison test (December '72). The air cleaner hardware and hosing subtract five horsepower at the rear wheel. Inside, the Sport has different pistons which bump up the compression to 9.5:1. Valve sizes and cam timing, however, remain identical to the GT-750.

Those who have read *Cycle*'s Ducati 750 road test (October '72) and the Superbike Comparison Test already know that the *Cycle* staff really liked the GT-750, so our normal curiosity was laced with genuine enthusiasm when we had a chance to sample the Sport in Montreal, Quebec—courtesy of Norstar Trading Company which imports Ducati (as well as Laverda and Benelli) into Canada. Norstar had two Sports, and they were the only ones in North America at the time.

Visually, the Sport is a striking motorcycle in bright yellow with black detailing. The striping scheme is attractive only from certain angles; generally the black accenting seems too fussy and overwrought. The side covers, in both their form and detailing, are less successful than those found on the standard version. The bike, we think, looks better without the covers in place. Likewise, the engine's black side covers seem overly contrived. One could quarrel all day about the detail touches, but there is no argument about the basic form of the tank and seat: if you like the café racer idiom, the shapes and proportions of both components harmonize extremely well with the machine's mechanical layout. The seat has a tool compartment in the tail section; one

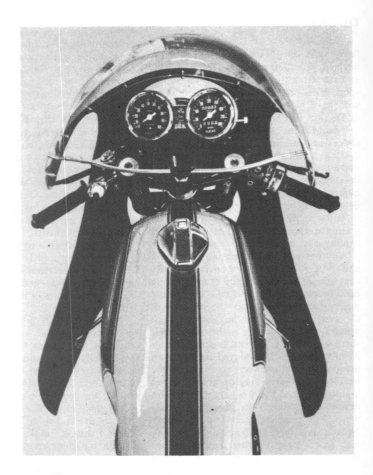

ONE FOR THE ROAD
DUCATI 750 SPORT

PHOTOGRAPHY: JIM MCGUIRE

Clip-ons and rear-sets intensify low-speed heaviness, but at speed the handling is great.

Rear-set peg folds up and stays outside the arc of the kickstart lever.

32mm Dell'Orto pumper carburetors (with rock guards) feed the V-twin.

simply unlocks the seat padding and lifts it up to gain entry to the compartment.

The seat padding and the riding position automatically disqualify the Sport for long-distance hauling. It's strictly a *Walter Mitty Speciale*, built for honking hard along back-country roads. At this task the machine is simply superb. The controls operate crisply, and there is a one-to-one feeling of correspondence between the rider and the machine parts which the controls operate. The clip-ons and rear-sets—combined with the excellent handling and Dunlop TT-100 tires—add to the fun of the countryside swervery. But the racer get-up intensifies the low-speed heaviness in the Ducati's handling. In short, stay out of town, and that's not such a bad idea anyway if John Law in your hometown takes exception to those 90 dB(A) notes pouring out the pipes.

The half-fairing keeps the air-blast off your midsection, but in so doing, there is not sufficient air pressure to pull the weight off your arms—which take quite a load with clip-on bars. The fairing has another kind of drawback for night riding. The headlight peers out through the plexiglass nose-window and this set-up isn't satisfactory for high-speed night work: the plexiglass weakens the beam, and there is a certain backsplash of light which never manages to get outside the fairing.

To the café racer, all of this may count for little, because the half-fairing really helps the appearance of the bike. The engine modifications add to the performance, particularly in the mid-range engine speeds. The Sport is not substantially stronger than a GT-750 *sans* air-cleaner plumbing; based on our Montreal experience, we would guess that a well-tuned Sport would deliver somewhere between a genuine 55 to 59 rear-wheel horsepower. Translated into road figures, this would suggest quarter-mile times around 12.7 and a maximum speed very near or slightly over 130 mph.

Since boosting the standard GT-750's output by 10% requires almost nothing in the way of tuning wizardry, after-market suppliers may soon offer performance pieces for the ordinary model. It appears that the new Ducati engine will return substantial increases in power for a moderate amount of fiddling. If accessory manufacturers jump into the act with some quality items, there's every reason to believe that the GT-750 owner could make his standard model as strong, perhaps stronger, than the factory's yellow-and-black special. And if you like the Sport's lines, before too long it may be possible to buy fiberglass work patterned after the general form and shape of the Sport, and doing it yourself could be cheaper.

One thing is certain. The Sport is not just a show piece. It is quickly establishing its credentials as a first-rate flyer. The particular bike which *Cycle* sampled was entered in the Five Hour Production Race at Mosport, Canada. The machine breezed through the distance with Richard White and Manuel Radbord in the saddle, held the opposition 50 seconds at bay, and collected the first place hardware. But not for long. After the race the bike became the center of a stormy protest. The protester claimed that the bike was not a production machine and therefore ineligible to compete—and to win. So the Sport was tossed out after the fact.

Meanwhile, White was cheerfully accepting orders. The price is a princely $2700 in Canada. This does not include the cost of race track lawyers and other quibblers. According to White, that's an optional extra. ◉

Mick Walker's 350 Ducati

A new racer for less than a CB500 road bike

IF you are hooked on motorcycles early, get involved in racing, and find that a regular fix of Manx GP atmosphere has imperceptibly become part of your basic needs, then you will need a suitable machine for the job. There is a certain group of riders who will forgo the whole short-circuit season just so that the Irish road races and the two non-TT Manx meetings can be supported fully. To this band of riders—who show a matured and impeccable taste, in my view—the full expense of the new season's Yamaha would not only be way out of sight financially, but also would not necessarily be the best tool for the job anyway. Reliability, good handling and robustness are at a greater premium than the last dozen or so unusable horsepower. The Aermacchi 250 and 350 used to cater for these people to perfection, but although it is still possible to order a 1973 model 350 Aermacchi this is not widely known, nor is the bike to many people's taste. The Aermacchi flavour is a bit different from most others, and there can be no other really good reason beyond the visible eagerness of the AM-HD factory to produce an Italian water-cooled Yamaha as fast as possible. This imminent threat of instant obsolescence (and the consequent danger of faded factory interest in the four-strokes of yesteryear as the new and fully competitive machines come in) might well have a lot to do with it. Nevertheless, a ride on the Lawton 402 Aermacchi is eagerly sought.

The other Italian firm which has consistently produced and supported racing four-strokes is Ducati. From the impeccable d.o.h.c. 125 twins to the latest 750 vee twins they have resolutely stuck to four-strokes, and the Ducati flavour is spiced with the classic bevel-drive overhead camshaft that is the *sine qua non* for a true Ducati four-stroke road machine.

Last season we saw Alan Dunscombe on one of the very rare 750 vee twins, but effectively there simply aren't any vee-twin Ducatis about: the Vic Camp 750 being the only one that I have even seen on either road or track, in spite of strenuous efforts to get at one to try for size, as I had considered buying one recently.

Mick Walker has stuck to the "classic" Ducatis, the s.o.h.c. singles that have given so many club riders their initial experience, and the learners at the Camp school their first taste of the track. I had a Mach I 250 for a season myself, and have been thoroughly in favour of the machine ever since.

When Hannah finally sold out his interest in Ducatis, Mick Walker dived in and bought up a huge amount of spares, engines, etc., to add to his already impressive stocks of things Ducati. It is this vast pile of bits that let him cautiously state that he has "nearly" complete cover for his racing engines now, and much the same for most of the road machines. Of course everyone likes to claim this, but from all reports it seems not unjustified in this case.

Mick has raced in the Manx GP, and has been actively supporting Ducati riders for some years now. Last year we tried his Production 24 Horas Ducati 250, and this year we tried his lightweight 350 racer. This 350 is the pilot machine for his current series of custom-built racers that he offers for sale. Dave Arnold has been racing it with considerable success recently, and Dave Street will be using it in future. The machine is basically a specially designed frame with a mildly tuned 350 road engine.

The history of this bike is interesting: originally it was built up for a customer as a special order. Saxon were asked to build up the frame and the whole bike was set up for this order. Things didn't quite work out, and after passing through the hands of John Blunt, who works at Mick Walker's shop, Mick bought it back and set it up for Dave Arnold. When I tried it it had just been fitted with a Fontana brake from another racer just bought into the shop, and which had useless linings fitted within. The normal brake is either a Grimeca or a Spondon unit. The specification to which it is now built could have been achieved eight years ago, and before the Yamahas were dominant. This is rather sad, because the performance now obtained puts the Ducati on a par with the Aermacchis, Nortons and AJS machines that set the pace in those days. Even now, on Cadwell Park style circuits it can put up brave performances: Dave Arnold found that he could match Paul Cott's 354 TR2 Yamaha in one of the early meetings this year.

The test was carried out at Silverstone under overcast skies, and with damp patches on the track. Mick arrived with entourage and unloaded the machine, only to find that he had no petrol. This slight oversight allowed a quick photographic session before moving out onto the track.

At first sight the Ducati looked slim, low and long: the sort of bike that you sat in rather than on top of, and everything was visibly tucked neatly away to make the most of the slim cross-section and to further the ground clearance. The exhaust system is slung underneath the bike, and emerges on the side opposed to the final drive sprocket. This certainly helps to improve ground clearance, as I found on my own machine.

The front forks are Ceriani, the front brake a Fontana . . . the frame is of the spine variety

The test was carried out at Silverstone

The front forks are Ceriani, the front brake a Fontana, and the rear brake Ducati. The frame is of the spine variety with a suspended engine. T45 tubing is used, and the number of tubes, sections and junctions look a little untidy on this pilot model, though it is cleaned up on the current production versions. The steering-head bearings are taper rollers, and needle rollers are used to support the swinging arm. The steering head is effectively supported by the multiple tube rails at top and bottom of the steering stem tube, and the overall torsional rigidity should be considerably greater than that of the rather flexible standard Ducati frame layout.

Riding the Ducati is easy: the single-cylinder engine is slung low enough to keep the centre of gravity right down, in spite of the heavy cylinder-head assembly. The slim engine casings allow a low position in the frame, limited only by the depth of the oil sump under the engine, and the need to sling the exhaust pipe underneath that sump. The front and rear suspension units are well set up for someone a shade lighter than my 13 stone, and so I got a comfortable ride. The Saxon tank is slim and low, a neat piece of aluminium construction, and the seat is canted slightly to locate the rider naturally. The hump at the rear in conjunction with this cant does the job. My six-foot frame fitted easily into the riding position. The Ducati is not overlong, so it must be the slimness between the knees allows me to feel so at ease on it.

I watched John Blunt trundle the Du round for a while to check that all was v

The immediate impression was of an ex lent gear-change. My old Mach I had a ra heavy change which cried out for impr ment. This gearbox had a very light acti with a fairly long throw. The modification the gearbox are one of Mick Walker's improvements to the machine. The throw the change action is adjustable, and the l throw merely reflected the preference of previous rider.

The Ducati was not that easy to get goi however: the 38mm carburettor requires f treatment, and my usual pussyfooting in use of the throttle. The engine died o thoroughly soaked plug, and only with help of a hearty push from John Blunt afte

The mildly tuned engine is suspended from a T45 tubing frame, with the exhaust system slung underneath

ick Walker's
0 Ducati . . .

plug had been put in could I get it going. ngely this did not happen again. Every I started it it barked immediately into life. large carburettor really made its presence up to 6,000 r.p.m., when the engine stopped plaining and got down to work in earnest. ve a suspicion that a Gardner carburettor d well give an increased, usable spread of er, as it is very well suited to the smaller—c.c.—version of the Ducati motor. The ati had close ratio gears fitted, and it was very simple to keep the engine within the 6-8,500 r.p.m. range allowed with the five gears. The initial lap was spent at very low speeds simply to get the feel of the bike. It was solidly responsive with that hunched-down feeling that one so often finds with really well set up small machines. The tyres seem to dig in on sharp bends, and the feel of the bike on fast bends is one of pushing into solid grip. Many 125s have the feel of being on tiptoe as they scrabble round corners, and some really powerful 350s and larger machines feel continuously mutinous, and on the point of waving front or back wheels in a sharp and catastrophic arc at any instant. The TD3 Yamaha and this Ducati both fall into that intermediate class where the tyre sections are large enough to overcome the babybike twitches and yet the power output is comfortably low enough to allow one to hurl them into corners with the throttle on the stop. I hasten to add that the TD3 in question was definitely on an off day, and seized up on the next outing with sundry crank maladies, otherwise the power could well have pushed it up into the wheel-waving bracket again.

The brakes were quite horrid. Mick had always wanted to fit a Fontana, and when one came his way he had put it in without relining it. The brakes were consequently quite useless and little constructive can be said about them. They have subsequently been relined and are now working as they should be, or so I am assured. The handling made up for it, though, and the accurate and precise steering was exemplary. It is interesting to reflect that the roadholding of many machines is no improvement on the standards prevalent a decade ago, but the steering seems to be better on almost all machines than would have been praised as exceptional only five or so years ago. Perhaps this is due to a better understanding of the importance of the torsional rigidity of frames and the need to pin down the steering stem in a constant position relative to the rest of the frame. Whatever the reason, it is always harder to tune a suspension system for all riders than to get the basic geometry correct at the design stage.

On one or two occasions the change from fourth gear was missed, and I was glad to hear that on inspection of the gearbox a fault was found, as the change action gave one full confidence until the disconcerting failure to engage.

The Ducati is light, and cannot weigh much more than 230 or 240 lb wet, and as the weight is kept low down the change from one direction to another is achieved with the minimum of effort. The engine braking on the overrun is strong, and the combination made the Ducati an extremely tolerant bike to ride. There was some vibration, but this was traced to the omission of an engine location bolt at the rear of the engine, which acts mainly to reduce the vibration as the engine is entirely located by the head attachment points and the upper rear engine bolt. On some machines it works quite the other way, as on a TD2 the omission of the same bolt will materially reduce the vibration felt by the rider.

In spite of the poor brake linings I soon was lapping at a fairly decent pace without any great effort, and in fact I was a bit surprised as I had been riding so easily that I had not thought that I had been going round so quickly for so little effort. This really summarizes the character of the Walker Ducati:

Outline Specification

340 c.c. Ducatti s.o.h.c.
Five-speed gearbox with close-ratio gears and modified selector mechanism. Special camshaft and racing valve springs.
35 or 38mm Dellorto carburettor with remote float.
Bosch 6v total loss battery-coil ignition system.
Spark plugs, Lodge RL49.
Oil, Shell Super M.
Borrani rims. Tyres: 2.75x18in front, and 3.00 × 18in rear on WM1 and WM2 rims.
Aluminium tank, fibreglass seat, mudguard and fairing. Spine frame: taper roller steering head races, needle-roller swinging arm bearings.
Choice of Spondon or Grimeca 9in front brakes.
Metal Profiles front forks.
Krober or Smiths rev-meter.
Girling rear-suspension units.
Cable-actuated Ducati rear brake.
Tomaselli clip-ons and levers.
Dry weight, 220 lb; approximately 230 lb wet.
Total cost of complete machine with sprocket set and including VAT, £685.

The extra not in the list is a valuable intangible: the bike is sold by an enthusiastic dealer who in a short time has built up quite a name for his racing services to Ducati riders. The Ducati concessionaires must be very happy to have such an energetic protagonist for what is after all very much a limited market for their specialized line of machines. Of course the testing time is now coming up: Mick Walker recently opened his new shop in Wisbech, and one sees so many good small businesses built up on one man's energy and efforts, which then falter and lose momentum when they begin to grow. Motorcycling is fortunate in the number of such small concerns . . . it would be even better if there were more larger ones. Meanwhile we can be glad that enthusiasm for racing helps to bring people like Mick Walker, Roy Baldwin, Jim Mortimer, Geoff Monty and many others into the game of producing and selling well-tested components and complete machines. We'll be trying some of the Clubman equipment produced by the others mentioned shortly: I hope they all measure up to the standard of this one.

solid and stable, with a deceptively lazy style of progress: absolutely ideal for a beginner on any circuit, and excellently suited to the long road races in Ireland or the Isle of Man.

The 250 Ducati seems to make an equally good powerplant but the 450 has never been outstandingly successful in the UK. Alan Dunscombe had some success with a 450, but Dave Forrester who next used it never really got very far in spite of his clear ability shown with a 250 c.c. Desmo Ducati. I'd really like to see this lightweight Walker set-up used for both 250 and 450 versions; after all, these two engines are merely bored out and sleeved versions of the 350.

The reliability and robustness of the 350 are difficult to demonstrate. The record of this prototype is as follows. Forty races in 1972 season, with one retirement due to a broken rocker shaft—an ailment that I too suffered in 1972 and traced to a faulty batch of rockers. With the racing and testing mileage stacked up to date, about 2,000 miles have been covered, and this plus one clutch plate and a condenser is the sum total of the engine parts required. The engine was stripped and checked at the end of 1972 and the gear-change fault is the first thing to show up since the rocker shaft went.

The engine has a number of modified parts to improve both performance and reliability. The con-rod is a special steel one, but as the camshaft is designed to produce increased power at below 9,000 r.p.m., there is no need to go to the titanium extreme, nor are the expensive special coil springs required to control the valves, which make do with racing specification hairpin springs. The use of coil springs, such as those made by Fath, allows the use of much higher r.p.m., and indeed produces more power. Unfortunately these springs can accelerate the rate of wear of the camshaft and the use of 10,500 r.p.m. or so has a similar effect on the life of the con-rod. Both these effects are different aspects of the eternal war between cost and performance. The Walker Ducati is designed to place the emphasis on reliability. The valve sizes are 35mm exhaust and 39mm inlet, and the cost of replacement valves is £2. The other parts are also fairly cheap and readily available : £1.75 for main bearings, and £5.90 for a complete set of 13 clutch plates. The plug used is an RL49 Lodge, which seems to be good for about 12 races before it needs replacement. The crank is balanced, and a forged piston is used.

The production 350 Walker Ducati is offered with a 35mm carburettor, and I would expect it to be even a little nicer with one such as Dellorto fitted, as the 38mm instrument is definitely a bit too big for the most tractable and fuss-free performance. The usual equipment required to get a racing bike together is familiar : Smiths or Krober rev-meters—there is a Krober on the prototype, Tomaselli lever and clip-ons, Girling rear units. Up until recently Mick Walker would provide a completely flexible service for the Lightweight : effectively you name it, you get it. Now that the number of orders are climbing it is beginning to get rather harder to carry out quite such a range of variations, so an outline specification is now being set out. Frame kits and engine uprating are of course stand alone services, and there are several further engine improvements under investigation at the moment. Presently helical gears are used to transmit the power at the primary stage, straight-cut gears are now being produced to try out. Camshaft designs are under continual study and experiment.

There are a few things that I would try to see if they would improve the Lightweight : a Gardner carburettor, 12 volt ignition, and possibly a Boyer ignition system to make the Ducati almost maintenance free. The strong rectangular section swinging arm needs no alteration, and the snail cam adjustment system is fine. The frame could perhaps be tidied up a bit, but that is already in hand. The seat, tank, footrest and clip-on arrangements are well set out as they are. The Ducati seems to add up to the sort of machine that a beginner would get lots of good experience with quickly, without the need for fine tuning and preventative maintenance, with enough performance to give him a good season to get the full measure of. There are few better ways of going racing than getting a new machine built specifically for the job by someone or some factory with solid racing experience. There are several such machines available now for less than seven hundred pounds. There are a production bike and a two-stroke racer, but the Walker Ducati seems to be the best four-stroke in this class.

There are few better ways of going racing than getting a new machine built specifically for the job by someone with solid racing experience . . .

Continued from page 131

shipped the hub out to him. The job wasn't easy, but Tim did it anyway, and did it successfully.

Hi-Lustre Metal Finishers in Lindenhurst, New York, handled the rechroming. What they received were scaggy and pitted pieces, what came back were re-chromed parts of better-than-average quality, and certainly better quality than the original factory chromework, which was (in the early 1960s) marginal. Hi-Lustre earned one big distinction—they lost a tank air-breather fitting, and the loss put me in a snit for a month.

I wanted to top off the restoration with a back-to-original paint scheme, and I wanted the paintwork perfect. Bob Brown, that cheerful troll who masquerades during daylight hours as the Editor of *Car and Driver*, guaranteed me that Gary's Auto Body in Sea Cliff, New York would do flawless work. Jess Thomas agreed. They both promised me that my bank account would get fragged. They were right. Herbert Gary brooks no compromises. You pay for his talent, and you get perfect work. It's that simple, and it was fine by me.

So how much did the re-restoration cost? I honestly don't know because I lost interest in counting the dollars when the total passed into the four-figure range. I confess to a certain morbid curiosity about the grand total, but only at the outset of the project. About halfway through the re-restoration, I decided that I really didn't want to know. The whole idea of cost became a moot question—the resurrection of my Formula 3 was just something that I *had* to do. Of course pricetags are important, but the re-creation of the 175 was an act of passion. At bottom, it had nothing to do with any sort of bookkeeping reality of dollars and cents. So what is the bike worth now? Everything and nothing. It's worth just as much as a lovely daydream or a pleasant memory.

For many restorers, projects are never quite completed. Along the way, compromises made out of necessity or cost force the use of near-original parts. This situation leaves the restorer looking for original pieces for several years, after the bike has been reassembled. My Ducati 175 required several compromises. The rear fender is not exactly like the original; the factory coating has been bead-blasted off the hubs and brakes; Baroni shocks replace the Marzocchi units which were destroyed in the fire; rims and spokes differ from the 1961 components. And all the road equipment remains missing. Although the Ducati Formula bikes brought into the United States had "street" equipment, those items weren't related to the function of the machines. If I had the street rigging, I don't think I'd mount it. Somehow I sense that the Italian who drilled all those holes inside the engine would understand.

—*Phil Schilling*

NEW TRICKS FROM ITALY

DUCATI'S 1973 DESMO RACER

BY BRUNO DE PRATO

● On April 23, 1972, Ducati smoked off the field at the first Imola 200 with a strong one-two. It was a tremendous show. The bikes went beautifully, strong and flawless, although they had been put together from stock components only a week before. Morale at Ducati was sky high. They felt they could conquer the world and win any race.

Only one man did not join the bandwagon. The man, as one might expect, was Dr. Fabio Taglioni, who had been responsible for the Imola hit. He knew it was no time for resting on laurels; Ducati had met only the European share of the opposition; the strongest 750 racers were in the hands of the American teams. Taglioni had seen this American equipment at Daytona. The bikes were so fast around the

The racing frame's swingarm allows rider to dial in three different wheelbases.

The new frame is lower and lighter, and the engine is more compact than the 1972 bike.

The new racer weighs in at about 325 lbs. Wheelbase is eleven cm shorter than '72 racer.

The racing department (minus Dr. Taglioni) with one of the three new Imola bikes.

oval that he could not read the numbers on the fairings. Imola is not Daytona; the Italian circuit is extremely demanding on the frames and suspensions. Maybe the superior handling of the Ducatis would have given them an edge even on the super-fast Japanese threes at Imola or similar tracks in 1972, but at Daytona the Ducatis would have been outgunned.

The Ducati 750 that won Imola '72 had so much room left for improvement that it would have been a shame stopping the development of a promising design, which was still far from its ultimate potential. It was not just a matter of getting a lighter and shorter frame in place of the stock unit Ducati had used. Actually the whole bike was simply too stock to be competitive in 1973 against the top teams in the world. Imola '73 had just proved that the basic Ducati 750 was a brilliant motorcycle, but 1973 required a real racer. Taglioni already had everything on his drawing board and in his mind. He wanted to show that the quest for multi-cylinders was just about nonsense, that his twin—the ultimate twin—could match the two-stroke missiles' flat-out speed, and out-handle and out-accelerate them through corners. Taglioni was stalking for Daytona '73. This exciting challenge, sadly, did not turn into reality because it was decided that the 1972 Imola machines should be taken around the world in '72 and raced wherever a Ducati distributor morale needed some boosting. So the limited manpower of Ducati's race department was strung out in this promotional campaign, which failed to produce anything matching the Imola thunderclap.

Meanwhile, back in Bologna, Dr. Taglioni was refining the project for his ultimate 90-degree V-twin. Bore/stroke ratio had to be more radically oversquare: from 80mm x 74.4mm to 86mm x 64.5mm, a real ultra-short stroke in order to exploit all the revving ability guaranteed by the excellent balance of the alternating masses in the Ducati's engine. With the 64.5mm stroke, the rev limit would presumably go from 9,200/9,500 rpm of the '72 machine to 10,200/10,500 rpm. The clean jump of 1,000 rpm forecast a power increase. The shorter stroke and shorter connecting rods also meant that the engine would be about one inch shorter and lower. Consequently, the L-shaped engine could have a much more compact frame built around it.

The new, radical oversquare bore/stroke ratio required a new head design. The stock heads feature valves with an included angle of 80 degrees which was too wide. To get an adequate compression ratio from those heads, in combination with the new bore/stroke dimensions, would have required pointed-top pistons with a consequent poor combustion chamber shape. Taglioni had been thinking of a new head design, more in line with contemporary practice for some time, but until 1972 his old heads had still proven effective. This would have been the occasion to retire them after about sixteen years of service—a design which started with the now long-gone 175 models. The 1973 heads had to have the valves set about 60 degrees apart so that very high compression ratios could be attained with almost flat-top pistons.

The drawings concerning these basic modifications were completed by Taglioni about mid-May 1972; by July the head castings were ready. Yet until the race season was over, practically November 1972, no one in the race department had the time even to dust them.

By November it was already too late for Daytona '73, but at least Ducati could get the bikes properly honed for Imola. Even that seemed questionable. 1972 was the year to renew the metal workers' contract

'73 engine: dry clutch, 86-x-64.5mm, 60° valve angles, desmo valve gear and 5-figure revs.

Carburetion remains the same as last year: two 40mm Dell'Orto pumper carburetors.

The new frame provides for chain adjustment via eccentrics in swingarm bosses.

in Italy, and a contract renewal in Italy is an affair which drags on for months, crippling all activities with an endless sequence of short, on-and-off strikes. By March 1973 the contract had still not been signed, and Taglioni's new heads remained raw castings. Then, with about one month left to Imola, the situation cleared up. The contract was signed and strikes were over, and Taglioni had full command of the race department again.

One month to set up, tune and test a practically all-new racer is ridiculous, but Wizard Taglioni is used to playing hurry-up catch-up. While special frames were being fabricated from chrome-moly tubing by a specialized shop, the engine was taking shape quickly. Luckily the bore of the new 750 was the same as the 450 that Ducati also raced in 1970, so race pistons were readily available. On the other hand, there was the extremely time-consuming preparation of the new heads, desmo heads of course, which required new camshafts and rockers. Everything had to be carved out of solid billets because there were no dies. Slowly the top of the new engine took shape.

For the bottom end, things were much more straightforward. A stock crank assembly was retained, but with con rods which were shorter than the standard 750. A dry clutch went in to replace the oil-bath unit. It saved some weight with its aluminum housing and the new clutch gave lighter action at the clutch lever.

With less than two weeks left, the first engine went on the bench. Everybody expected that the new engine, compared with the old one, would gain something at the top and lose something at the bottom and at mid-range. Instead, first readings showed that torque curve was extremely flat and that power was very strong right from 4,000 to 5,000 rpm, but then dropped (against all expectations) past 9,200 rpm. Immediately ports were enlarged and, in that process, slightly reshaped to allow better breathing. Low and mid-range power did not seem to suffer at all, while a couple hundred revs were gained at the top. The power increase over the 1972 engine was modest though the power spread was much improved. Everyone was disappointed. Despite very efficient port shapes, it looked like the engine was not able to breathe beyond 9,500 rpm. Taglioni decided to concentrate on the preparation of the whole machine and forget the dyno, for the moment at least. The short-stroke engine did fit into the new frame beautifully. The frame structure was basically the same as the production item, but eleven centimeters had been cut off the wheelbase.

The bike looked incredibly compact and mean the day it was brought to Modena to be tested by "old man" Spaggiari. After a few warm-up laps Spaggiari came in reporting some front end wag. It was decided to put on a new Marzocchi center-axle fork in place of the offset axle unit, but retaining the almost flat tri-clamps of the old unit. That would have meant an enormous trail increase, but at that moment there was no time for other attempts. Despite all speculations, the bike remained easy to handle and the wag had completely gone. When the engine got loose, Spaggiari began to push. It did not take much time until he knocked a clean two seconds off its previous 750 lap record. The new time, 59 seconds flat, was also the absolute record, .3 seconds faster than Agostini's previous best lap on the MV 500 three.

Spaggiari reported that the engine pulled strongly beyond 10,400, which did not sound right until an explanation was found about the engine's inability to rev beyond a certain limit while it was on the bench. Ducati's dyno is placed in a sort of cabin with brick walls which have no sound deadening capacity. A 750 engine revving at 9,500 created such a resonance on the test stand that the carburetor floats were completely upset and consequently the engine was either starving or was flooded. Until Ducati's new dyno room is ready, nobody will ever know how many horses the new engine pumps out.

The new machine was greatly improved in all departments over the old one. Acceleration was reported as blinding, road holding was extremely sure-footed and handling was showing the positive effects of the much shorter wheelbase and of the considerably reduced weight, which was down from 390 pounds to 325. The week

before Imola, all efforts were concentrated in putting together the other two bikes for Mick Grant and Swiss revelation Bruno Kneuhbueler. Further sharpening of the new machine was impossible and everyone had to be content with what they had in the first instance. The last engine was ready Thursday, April 12; the race was Sunday, April 15, so the engine had to be run-in during qualifications. It was Spaggiari's bike engine, and he finished in second place.

Kneuhbueler had the bike that had gone through all the tests, with more than 300 miles on the engine; he was a little concerned about the engine stamina, since he lapped Imola at 1.49.1, three seconds faster than last year Spaggiari's and Smart's record and only .2 seconds short of Saarinen's new lap record. The young Swiss has not been particularly lucky, while lying second and closing on Saarinen, he was knocked by a slower rider and broke his left-hand thumb. Mick Grant burned out the clutch of the third team bike at the start of the first heat, made the second with a practically fresh bike, but he did not feel like sticking his neck out for nothing, so he motored along lapping steadily at 1.54.0, an almost winning pace last year.

The engines ran incredibly steady and even. Though the Ducatis are not yet a match for the Suzukis as pure speed, they are a hell-for-strong 750s. This year Ducati will race 750s at the Bol d'Or only. Strikes are not expected, so for once Taglioni and his men should have enough time to get the best out of the present machine. And so, once again, Ducati may just have another winner. ◉

Italians also give us Ferrari, Lamborghini, and Maserati.

DUCATI

Type Usage	Displ.	Engine Type	Bore & Stroke	HP@RPM	Comp. Ratio	Carburetion	Ignition	Gears	Wheelbase	Dry Weight	Aprox. Speed	Aprox. Price

350

Brand new in America this year is one of Italy's most proven and popular sporting lightweights—the OHC Ducati 350 street single. Last year's 450 enduro/playbike has been dropped in favor of this fine handling, reliable thumper, which is capable of more than just a short off-road jaunt if the opportunity arises. A three-ring forged piston, roller bearing big-end and cast-iron liner join an alloy cylinder and head. Electronic ignition is new for '73. Aluminum rims, 2.3-gallon tank and Pirelli tires round out specs.

| Street | 340cc | 4/S ohc Single | 76x75mm | Max.@7500 | 9.5:1 | 1,30mm, Amal | Motoplat electronic | 5 | 55" | 295 lbs. | 90 mph | $949 |

750

If you are a motorcycle purist and have the money, don't buy anything before putting 50 miles on a Ducati 750. It will probably take even less time than 50 miles to convince you it is probably the finest motorcycle in the world. Four-out-of-five *Cycle* magazine road testers choose the Duke as the bike they would most want to own—the fifth selects another Italian, the 1000cc Laverda Three. The Ducati is smooth, fast and full of torque. Unfortunately it is a bit too loud, fairly scarce and as yet unproven as to reliability.

| Street | 748cc | 4/S ohc V-Twin | 80x74.4mm | Max.@7800 | 8.5:1 | 2,30mm, Amal | Bat. & Coil | 5 | 59" | 440 lbs. | 115 mph | $1964 |

"DESMO" DUCATI R/T 450 Road and Track

Just about everything any motorcyclist ever wanted in a true sports model has been incorporated in this meticulously assembled, engineered and styled DUCATI...
See Your Nearest Dealer

BERLINER MOTOR CORPORATION
Sole U.S. Distributors
Railroad Street & Plant Road
Hasbrouck Heights, New Jersey 07604

CIRCLE NO. 28 ON READER SERVICE PAGE

If you're ready, climb aboard.

Announcing the Ducati 750 twin. A brand new heritage for motorcycle enthusiasts... if you're ready for a really unique experience.

With the introduction of the 750, Ducati has not produced "just another twin". It has developed an unequalled blend of performance, handling and durability in a truly vibration-free ride that must be felt to be believed.

The 750 incorporates a 90-degree, longitudinal V-twin engine with a total cross-section no larger than most single-cylinder models, allowing a perfect alternating weight balance and maximum cooling for longer engine life; an open, double-cradle steel frame with the engine crankcase acting as the bottom frame member; constant-mesh, five-speed gearbox; twin carbs; hydraulic front disc brake; directional signals; new, modern instrument panel; specially designed and race tested front fork and suspension for pinpoint handling; and head-turning styling.

You owe it to yourself to test ride the 750 today at your nearest authorized Ducati dealer. You may never want to sit astride the competition again.

SEE THESE OTHER EXCITING DUCATIS AT YOUR DEALER NOW

BERLINER MOTOR CORP./Hasbrouck Heights, N.J. 07604/Sole U.S. Distributor
A Berliner Group Member/Norton•AJS•Ducati•Moto Guzzi•BeBe•Metzeler Tires

CIRCLE NO. 57 ON READER SERVICE PAGE

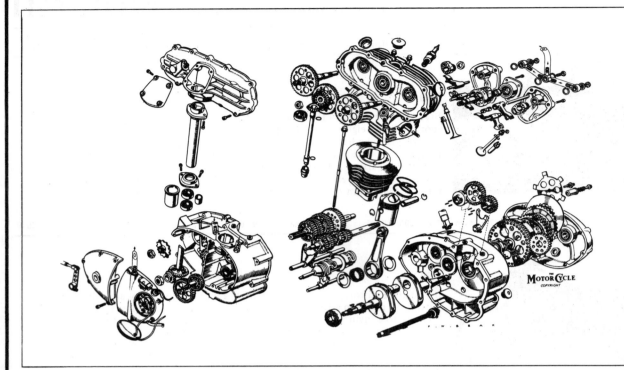

This exploded drawing of the 125cc desmodromic engine shows the details of a truly great classic racing engine.

BY GEOFFREY WOOD

IN DISCUSSING many of the world's motorcycles we speak of their great days in the past when the marque produced scintillating mounts for the more sporting minded rider, or when they added illustrious chapters to the colorful saga of motor racing. When we speak of Ducati, we speak of the present, because the Ducati is truly one of the most modern of all the motorcycles being produced today.

The story of this Italian masterpiece begins only a few years ago in 1950 when the Bologna factory was rebuilt after the devastation of the war. Previously, the company had never produced motorcycles, but it was decided they would enter the motorbike field because post-war Italy needed personal transportation at a price that the rather poor populace could afford.

With the capital stock partly owned by the Italian government and the Vatican, the company had the financial resources to launch a line of 48cc and 65cc motorbikes and a 175cc scooter. These models proved to be well-designed, peppy little mounts, and the company was on its way to success.

As post-war Italy flexed its economic muscles, the standard of living slowly increased, and, quite naturally, the Ducati concern grew right along with it. In 1954, the factory hired Ing. Fabio Taglioni to head up the design work, and this move proved to be an excellent one as Taglioni's genius was soon manifest.

By 1955, the Ducati was rapidly becoming a best seller in Italy, and the inexpensive little 65TS model was probably the star of the range. This 65cc four-stroke model produced 2.5 hp at 5,600 rpm for a maximum speed of about 44 mph. The suspension system was very advanced for those days, with a swinging-arm on the rear and an orthodox telescopic fork on the front. A comfortable dual seat was fitted, and, to satisfy the Italian desire to be a "racer," a small racing type windscreen was mounted over the handlebar. This windscreen bit seems very strange on a 44 mph motorbike to anyone except an Italian, but it made the 65TS a "sports" mount and this must have been important to the Latins, because the motorcycle sold in great numbers.

In 1956, the more affluent natives had their eyes on the 98S model, which offered more spirited performance than the 65cc Ducati. This new 98cc ohv motorbike had a more potent powerplant, producing something like 6.5 hp at 7,000 rpm for a maximum speed of 56 mph. The 98 model looked considerably more rugged than the smaller edition, and it had a heavier "open" type frame along with larger brakes to match the performance. The front tire was a 2.50 x 17 inch and the rear was a 2.75 x 17.

Still catering to the Latin desire for a "racer," the 98S had a small handlebar fairing complete with a tiny racing windscreen. With low handlebars, the mount was supposed to look very fast, at least to the Italians, and this rather humorous accessory on a 56-mph motorbike helped sell the 98S model in goodly numbers.

Ing. Taglioni had not been hired just to design roadsters. His great passion in life was in fire-breathing racing engines. Hard at work on his brainchild, Taglioni was destined to produce a truly great racing engine — and on a meager budget at that.

It was not until the 1956 Swedish Grand Prix that the new racer appeared, but the impact of the design was tremendous, as it set the whole motorcycle racing world to talking. The engine displaced 125cc and featured a desmodromic cylinder head that

DUCATI HISTORY

Currently available only in Europe, the 250cc and 350cc Ducati road racers feature a double twin-leading-shoe front brake and a special frame. The engine is basically standard except for cams, valves and carburetion. It produces 34 hp at 8,500 rpm from the 250, and 39.5 hp at 8,000 rpm from the 350.

llowed fantastically high revs with no fear of valve float. Desmodromics, for those of you who are unfamiliar with the term, means positive valve control, and in the Ducati this meant simply that there were a pair of cams to push open the valves and another pair of cams to pull them shut. Valve springs, therefore, were dispensed with, and this positive mechanical process meant that valve float was eliminated right along with the dangers of "holing" a piston or failing the valve gear through overstressing the components. At any rate, Degli Antoni astounded the racing world by riding the "desmo" machine to victory over its more established rivals in its first race. Then, just to seal the company's reputation, the Ducati team of Fargas-Ralachs rode a 125cc model into first place in its class in the Barcelona 24-hour Grand Prix d'Endurance — a true test of stamina and reliability at speed.

The early success of the marque declined a bit during 1957, when the company concentrated its resources on the introduction of the new 175cc ohc "T" and "S" (touring and sport) models, and these were followed in 1958 by the 100 and 125cc "Sport" models. By then, Ducati's goal was obvious; they were out to expand their sales to all corners of the world. To do this meant two things: they would have to produce a line of really good motorcycles, and then they would have to get the message across to riders all over the world. The new ohc Ducatis certainly fulfilled the first requirement, and for the second it was decided, and wisely so, to conduct an aggressive campaign on the world's grand prix tracks. This racing effort would have to be conducted on a slim budget, though, because the company was still rather small and the funds were not too plentiful.

As far as world-wide sales were concerned, Ducati had a really fine machine in the new 175cc ohc model. The 175cc alloy engine, with a 62mm bore and a 58.8mm stroke, produced 11 hp at 7,500 rpm for a maximum speed of 68 mph. Carburetion was handled by a 22mm Dellorto, and the valves were inclined at 80 degrees to each other. The compression ratio was a modest 7:1.

The engine's lower end featured a rugged ball and roller bearing assembly, and the crank and transmission cases were cast in one clean-looking unit. The gearbox had four speeds. This engine-gearbox case was mounted in an exceptionally well designed frame which used a swinging-arm rear suspension and a telescopic front fork of outstanding quality. The brakes were very large and were housed in beautifully finished alloy hubs. The whole bike had an exceptional finish, and the engine had Allen-type screws instead of the slotted screws on the covers.

These 175cc models were followed by the 100 and 125cc sports models in 1958, which were produced more for domestic consumption. The little 100 Sport model proved very popular with the Italians because it provided a speedy little tourer to very sporty specifications at a cost factor that they could afford. Similar to the 175cc models except in size, the engine had a 49mm bore and a 52mm stroke. A compression ratio of 9 to 1 was used along with an 18mm carburetor, and the little lunger pumped out 8 hp at 8,500 rpm. Top speed was listed as 65 mph.

During 1957, the factory spent their time in developing the 125cc desmodromic engine and improving their standard production range. Their racing successes, quite naturally, declined a bit. Just to keep their name alive and to prove that 1956 was no fluke, the factory sent Bruno Spaggiari and Alberto Gandossi to Spain for the 24-hour Barcelona event. Once again the 125cc ohc proved to be reliable as the team garnered first place at record speed in the 125cc class.

In the winter of 1957-58, the factory shops were very busy. The Ducati management had aggressively set up an extensive dealer network all over the world, and retailers in the European countries, North and South America, the Orient, and Australia, all had this new line of ohc singles to sell. The management promised these new dealers all over the world that, by the end of the year, Ducati would be renowned as a great motorcycle.

To get the Ducati recognized meant just one thing — successful participation in world championship racing. To this end the factory dedicated itself in the spring of 1958. It would take some determined work to show the world the brilliance of Ing. Taglioni's designs. Still a relatively small company, Ducati did not have the finances to hire the very top riders or conduct a massive racing campaign. Still, they had the desmodromic engine, and, by choosing their battlegrounds, they just might steal the whole show.

The very summit of international racing is the Isle of Man TT, and the marque decided to start the season there. Handicapped by not having riders with a great deal of experience on the island course, the Ducati team didn't really expect to defeat Carlo Ubbiali and his MV Agusta. The team did put up a courageous battle, however, and Romolo Ferri, Dave Chadwick, and Sammy Miller took second, third, and fourth places.

The next battleground was the Dutch Grand Prix at Assen, and when the little 125s were pushed off the grid, it was Alberto Gandossi and Luigi Taveri who led the first lap. After a fantastic battle it was Ubbiali again, though, with Taveri, Gandossi, and Chadwick taking second, fourth, and fifth places. Taveri was beaten by just a few yards, though, and he did have the honor of making a record lap.

Then followed the Belgian GP at Spa, and this was where the tide turned. The previous races had been on the slower, more twisty courses where Ubbiali's brilliant riding gave him an advantage, but now the season was to enter the faster courses where horsepower would show. The result was a smashing victory for Ducati, with Gandossi, Ferri, Chadwick, and Taveri taking first, second, fourth and sixth places.

The next race was the German event at the Nurembergring and once again the Ducatis streaked into the lead. Displaying superior speed over the world champions, the Ducati riders steadily pulled away. Then, disaster struck. Ferri crashed, Taveri's engine went off song, and Gandossi's model broke down.

Then followed the Swedish Grand Prix at Hedamora, and Gandossi and Taveri

This racy-looking 100cc Ducati was produced in 1958 — a popular mount for the racing-conscious Italians. Of high quality and excellent design, the ohc engine was good for 65 mph.

The European version of the Diana is the Mach I, claimed to be the fastest standard production 250 in the world with a top speed of 106 mph. This model is available with either high or low handlebars.

once again were the victors, with Ubbiali taking a bad drubbing into third place. Next was the fast Ulster event, only that year the race was run in a driving rain. Gandossi rode like a tiger that day and, with two laps to go, he held Ubbiali at bay — clocking 107.9 mph on the Leathemstown Straight. Then, on the hairpin, Gandossi lost the model on the slippery road. He was able to remount and finish fourth, with Taveri and Chadwick in second and third; but with his crash went his last chance for the world championship.

The final event of the 1958 season was at home on the fast Monza circuit. With great prestige at stake here, the two an-

The fabulous 250cc twin-cylinder desmodromic racer which Mike Hailwood used for many victories. Only one model was built, especially for Hailwood, and then the marque lost interest in racing.

The first of the ohc Ducatis was the 1957 175cc model. This one is shown in export trim for the U.S. market.

tagonists made great preparation for the race. This final chapter in a season of torrid racing was destined to be Ducati's, though, as the marque quite literally slaughtered the mighty MVs by taking the first five places. The final order was Bruno Spaggiari, Gandossi, Francesco Villa, Chadwick, and Taveri. All the Ducatis were desmo singles except Villa's mount, which had a new twin cylinder engine.

So the curtain fell on a courageous effort by a small factory — and it very nearly proved successful, too, as only Gandossi's spill in the waning laps of the Ulster Grand Prix stood between him and the championship. This racing campaign was eminently successful for what it was intended to do, however, and that goal was to put Ducati "on the map" all over the world. Almost overnight the marque became famous and sales sharply increased. The year of 1958, then, must be regarded as the turning point for Ducati. And just to add some frosting to the cake, the team of Mandolini-Maranghi once again took the 125cc class at the 24-hour Barcelona event to cap a really great year.

For 1959, Ducati planned a racing campaign that would keep them in the news, but it was neither as extensive nor expensive as the all-out effort of the previous year. Some remarkably good results were obtained, however, with outright victories by Ken Kavenagh in the 125cc Finnish Grand Prix and by Mike Hailwood in the Ulster event. Hailwood also garnered a third at the Isle of Man, and other Ducati riders took fourth, fifth, and sixth in the Ulster, as well as a third in the Monza classic. Then, once again, the Ducati team of Flores-Carrero won the 125cc class at the 24-hour Barcelona grind on a standard production 125cc Sport model.

In 1959, the standard production range was increased with the 85cc Bronco, which was derived from the older 98cc model, and this was later enlarged to a 125cc powerplant. During 1960, the ohc 175cc models were enlarged to 200cc, and in 1961 and 1962, the Piuma Sport and Falcon 48cc two-strokes were added to the range of motorcycles.

During 1960, Ducati racing successes took a sharp turn downward as the factory changed its competition policy. Racing in world championship classics had achieved what the company wanted in the way of publicity and sales, and so the factory turned its efforts toward improving production models and building racing bikes for the private owner. These production racers have achieved a great number of victories and placings in Europe, Argentina, Canada, the United States, Africa, Venezuela, Brazil, Uruguay, Ecuador, Chile, Australia, and the Orient, and in all these places the Ducati name has become synonymous with performance.

This knowledge gained in racing has been incorporated into the standard production models, and the 200cc ohc engine was enlarged to 250cc in 1962 and produced as a super sports model (Diana), a tourer (Monza), and a scrambler. In 1963, the factory began production of the 100cc Cadet and Mountaineer models, and then the 48cc and 100cc scooters. Lastly, in 1965, the 250 acquired a five-speed gearbox and new 160cc and 350cc ohc models made their debut.

The Diana MK III is probably the star of the current range, and this 250cc sportster is for the enthusiast who likes to play road racer on weekends. American road tests have obtained speeds of around 104 mph on this bike, which is truly remarkable performance. The fine handling and powerful brakes are also exceptionally good, and the Diana is probably about as

The 98cc ohv Ducati of 1956 was a popular motorbike in Italy. This clean-looking 56-mph roadster had a racing type fairing to please the sporting minded Italians.

close as a fellow can come to running a pukka road racer on the street. The fittings on the Diana are all rather sporting with low bars, an extra megaphone exhaust, tachometer, and rear-mounted footpegs and brake-gearshift levers as standard equipment for the machine.

For more sedate touring, the Monza model is the answer, with a lower performance engine and more comfortable accessories. For motocross fans there is the 250cc scrambles model. The 350cc Sebring and 160cc Monza Junior models complete the range.

In Europe, the Ducati line is slightly different than in the U.S., with the Mach I replacing the Diana model. This Mach I is claimed by Ducati to be the fastest standard production 250cc motorbike in the world, and few there are that argue with this claim. Several British road tests have attained 106 mph with the muffler intact — which is about two mph faster than stateside tests of the Diana with a megaphone. Chief differences between the Mach I and Diana are battery ignition, low bars, fuel tank, seat, tachometer drive, larger valves, and a different camshaft.

For pukka racing there are 250cc and 350cc racers available which offer excellent performance on the track. The Ducati racer features a massive double, twin-leading-shoe front brake and a full duplex cradle frame. The lower half of the engine is similar to the works racers, but the top half has the conventional single overhead camshaft. Power output is rated as 34 hp at 8,500 rpm on the 250cc engine and 39.5 hp at 8,000 rpm on the 350cc version. Both engines have twin spark ignition and five-speed gearboxes.

Ducati's aim in producing these racers for European sportsmen is to supply a racing bike that handles well, performs reasonably well, is reliable, and yet sells at a cost low enough so that almost anyone can afford it. A more powerful engine could, no doubt, be supplied, but the cost would soar and reliability would decline as a result.

Since the change in racing policy from exotic works desmodromic engines to production racers and clubman-type racing, Ducati has continued to establish an enviable reputation. In the classic 24-hour Barcelona race for production or prototype models, the team of Villa and Balboni took a 175cc model into a remarkable first place ahead of the works 600cc BMW in 1960. The next success was in 1962 when the Fargas-Rippa team took another first, this time in the 250cc class.

Then, in 1964, came perhaps the marque's greatest hour when the team of Bruno Spaggiari and Giuseppe Mandolini took a 285cc single ohc model into an outright win going the record distance of 2415 kilometers — the first time that 100 kph had been achieved for the 24 grueling hours. All this was done, mind you, against the finest big-engine teams that Europe could produce. Still displaying their stamina, stemming from good engineering, the Ducati team of Cere-Giovanardi again took a first place overall in the 1966 Imola Six-Hour Race, one of the big races counting for the Coup d' Endurance award. Behind the winning Ducati 250 were many big motors up to 750cc.

And so ends the story of the growth of Ducati. But, before we leave this speedy little Italian, let's go back to that fabulous tiny desmodromic engine and take a peek under the gas tank. A truly great classic in the annals of European motorsport, the 125cc Ducati is the only successful desmodromic motorcycle engine to have ever been raced. Produced by a brilliant engineer, the project was handicapped by a slim racing budget and a shortage of really top-flight riders. Nevertheless, it achieved some dramatic victories and added a chapter to motorsport history.

Taglioni first sketched out his desmodromic layout in 1948, but it was not until 1954, when he was hired by Ducati, that he could get down to the actual design work. In 1955, the first engine was assembled, and in 1956 it was entered in its first race in Sweden.

The little 125cc desmo engine had a bore and stroke of 55.25 x 52mm, and was the same as the marque's single and double overhead cam engines from the cylinder top downward. The two cams that opened the valves were located on spur gears in much the same manner as a double overhead camshaft engine. On the middle shaft was located the closing cams which had, of course, an inverted profile, and these closed the valves via short rocker arms. These rocker arms were attached to the valve stems through flanged collars located by split wire rings sprung into grooves in the stem. The valves were closed to within 0.012 inch of their seats, and the internal compression pressure then closed them onto the seats.

The valves themselves were inclined at an 80-degree angle, and the inlet and exhaust lifts were 8.1mm and 7.4mm, compared to 7.5mm and 7mm on the standard twin-cam Grand Prix model. The throat diameter of the valves was 31mm on the inlet and 27mm for the exhaust. Carburetor bore size was 22 or 23mm for use on tight, twisty courses which put a premium on acceleration, 27mm for typical grand prix courses, and 29mm for very fast circuits such as Monza or Spa-Francorchamps. The compression ratio was 10 to 1.

Power output was 19 hp at 12,500 rpm, compared to 16 hp at 12,000 rpm on the twin-cam Grand Prix model. The little desmo twin that Francesco Villa used in the 1958 Monza classic had a bore and stroke of 42.5 x 45mm and pumped out 22.5 hp at 14,000 rpm; but at the time the single was more raceworthy because of its wider torque range. The maximum speeds with dolphin fairings were 112 mph for the single and 118 mph for the twin.

The frames that these splendid little engines were mounted in were quite orthodox for their day and the handling was never reputed to be anything exceptional. Five- or six-speed gearboxes were used, depending upon the circuit, and twin-coil ignition was employed. The oil sump was contained in the crankcase and the oil lubricated both the engine and transmission. A castor-base oil of SAE 20 weight was used, and fuel consumption averaged about 45 mpg.

Another interesting Ducati produced in 1959 was a 250cc desmodromic twin that turned to 11,800 rpm. The factory produced just this one model, built especially for Mike Hailwood, and then lost interest in racing. On this fabulous twin, Hailwood set his homeland short circuit races alight as the screaming Ducati smashed many a race and lap record.

Today, the racing efforts of the marque are concentrated toward the European production machine racing where they do so well, and the improvement of their rather secretive experimental models. The regulations for the Coup d' Endurance racing trophy allow some modifications from standard production practice for experimentation, and Ducati has taken advantage of this. The result has often been some surprising victories against much larger engines, and the company has benefited immensely from the publicity.

All this participation in racing their production models has had its benefits to Ducati, and the knowledge gained is usually incorporated into the standard production range. With a constant technical improvement to their basic ohc design, the present-day Ducati range must be considered one of the most modern in the world.

From scooters and inexpensive lightweights on up to the fabulous 250 Diana MK III, here is a motorbike that appeals to the knowledgeable enthusiast. Quality of workmanship and race-bred design go hand in hand at Bologna. ∎

The new Ducati 750.

Announcing the Ducati 750 twin. A brand new heritage for motorcycle enthusiasts . . . if you're ready for a really unique experience.

With the introduction of the 750, Ducati has not produced "just another twin". It has developed an unequalled blend of performance, handling and durability in a truly vibration-free ride that must be felt to be believed.

The 750 incorporates a 90-degree, longitudinal V-twin engine with a total cross-section no larger than most single-cylinder models, allowing a perfect alternating weight balance and maximum cooling for longer engine life; an open, double-cradle steel frame with the engine crankcase acting as the bottom frame member; constant-mesh, five-speed gearbox; twin carbs; hydraulic front disc brake; directional signals; new, modern instrument panel; specially designed and race tested front fork and suspension for pinpoint handling; and head-turning styling.

You owe it to yourself to test ride the 750 today at your nearest authorized Ducati dealer.

BERLINER MOTOR CORP. / Hasbrouck Heights, N.J. 07604 / Sole U.S. Distributor
A Member of the Berliner Group: Norton, AJS, Moto Guzzi, Ducati, BeBe & Metzeler Tires

long playing single

How a neglected three-fifty Ducati was restored to mint condition.

A mountain of Ducati singles in Mick Walker's "Aladdin's Cave".

I REMEMBER walking around a locomotive "graveyard" towards the end of the era of steam. Mighty, rusted shapes loomed out of the mist from all directions — yet among the cluster of little tank locos and worn-out goods engines, a massive express passenger locomotive stood out in all its magnificence. Its nameplate had been unscrewed by souvenir-hunters; there was a large sack tied round its double chimney; the boiler was laced with rust and one of the massive connecting rods was missing — yet for all that its huge smoke-deflectors suggested power and beauty of motion and line. Even the graveyard could not remove its soul.

So it was when I first set eyes on my little Ducati Sebring. It had been standing forlorn and uncared-for in a gloomy backyard in York for countless months. The fuel tank had been roughly sprayed with an aerosol, as had the one remaining toolbox. White and green corrosion covered engine casings, finning, brake hubs and handlebar controls. The wheel rims were ruined, chrome plating hanging from them in curling-up strips. The silver-painted mudguards were going rusty and both were split, the exhaust pipe had lost all its chrome and the Silentium silencer had been painted matt-black to hide a rusty patch. The cables were all due for replacement and the clutch was so rough it would hardly move at all.

Yet through all the neglect, the character of the little three-fifty shone through. Its beauty and purpose could not be denied. It was a light, handleable sports bike through and through, put together almost like a mechanical sculpture by its Italian builders. It was pleading for someone to give it back its self-respect . . .

Well, I'm a silly sod really when it comes to things like that, and I duly parted with something in the region of one hundred green notes. Oh, yes — my conscience pricked me all right. "All that money. You're mad," I said to myself as I drove the van back to Lancaster, where I was living at the time. Then, lulled as the

The camshaft bevel gears have to be shimmed until this ground area lines up exactly.

Mutilated locknut and broken ball-end reflect previous owner's "tender loving care".

Although worn, this camshaft could be built up and re-profiled — we fitted a new one.

This stopper plate was well grooved — we were lucky, they usually break!

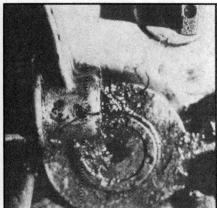
Disaster! — well almost, this crack in the crankcase had to be welded and bushed.

long playing single

Yorkshire scenery sped by on a beautiful spring evening, I started thinking in more positive ways.

Of course it looked a mess. It would have to be completely stripped down and refurbished, and the job would take ages. Where the hell would I get spares for an outmoded Duke anyway? Even so, a kind of relationship developed between me and that bike. I was determined to restore it to its former glory. It had been cruelly treated through no fault of its own. And I'd never owned an Italian sports bike before anyway.

Thus started the saga of the Sebring. For many long nights throughout the spring and summer of '78 I toiled away with paint-stripper and scraper (ie whichever screwdriver had the smoothest edge) and eventually removed all traces of paint and rust from the frame, mudguards, side-boxes, chainguard and so on. My hands blistered and tore with this crazy labour of love, but slowly and painstakingly, the bare metal came through.

I don't pretend to be able to paint bikes properly, so I took the frame and bits to a friend in Cheshire who paints lorries and vans. I decided to have everything except the fuel tank and side covers done in black, and the rest in red. These colours really look the part, coupled with polished alloy and the general lines of an Italian bike.

At the beginning of August, I moved to Peterborough and Motor Cycle Mechanics, and for a spell all work on the bike came to a halt. There was a new home to set up and a book to be written in the months leading up to Christmas. It was February again before I could get back to the bike.

When I'd first stripped the Duke down, I'd carefully removed the wiring harness, labelling each wire-end with insulating tape. The wiring was not in such a terrible condition as I'd thought, but even so there's nothing like playing safe, so a new one is on the way.

Apart from the inevitable nuts and bolts that were corroded solid (under mudguards, etc) I was delighted by the ease with which the Duke came apart. These Italian singles have a long and proud history reaching back into the racing era, and it's obvious that the Sebring had been constructed as a racer would: logically.

The wheel rims were beyond redemption. I sent the wheels to Blackburn wheel-builders Hacking and Kay for alloy rims and new spokes. The bill came to just over fifty quid. The brakes were in first-class condition, surprisingly. I know discs usually work best, but the little Duke's brakes are nevertheless very effective — and they look nicer than any disc on the market.

Now came the tricky bit — or so I thought: spares.

I knew of Mick Walker Motor Cycles right at the start, but Wisbech is a hell of a long way from Lancaster. Even so, I made the trip and gave them a list as long as my arm of bits and pieces I thought I'd need. Although I didn't hold out much hope, I was amazed by their concern for my out-of-production bike and the painstaking efforts they made to price every item I'd put down. They had just about everything, from gaskets to wiring harnesses, and the prices were very reasonable compared to Japanese stuff.

Some bits, like the front exhaust pipe, would have to be ordered from Italy, but it could be done, and even the missing side-cover, which was not in stock, could be replaced with a second-hand one from a broken up bike. At that stage, I'd never heard of Mick Walker's "Aladdin's Cave" or I'd have realised then that they could have found just about everything . . .

By the time I could make another trip to Wisbech, I'd joined MCM, a mere half-hour's journey from there, but then came the inevitable delay while I settled down to the new job and the book.

In mid-February, the urge came again. I rang Mick Walker's and arranged to have the motor completely stripped and rebuilt, ready for installation into the almost-completed remainder of the bike.

Before we started, I had a good look at the spares shelves and was surprised at the number of brand new original tanks, seats, mudguards, frames, wheels and other components for the whole range of Ducati singles. "There's more upstairs, you know," a voice said, and I found myself climbing a creaky ladder into "Aladdin's Cave". I couldn't believe it. It was a Duke enthusiast's paradise! All around were second-hand bits from a whole range of Duke singles. Some fuel tanks were still brand new, in their original wrappings. There were wheels, engines, frames, pipes, the lot! It was then that I KNEW my old bike would soon sing again.

I spent the rest of the day in their workshop with Dave Walker, watching my engine come apart. One crankcase half was found to have a serious slit around the clutch mechanism, probably caused by a broken chain during its previous ownership. So before any rebuilding progress could be made this had to be alloy-welded. At the same time I decided to have the outside of the engine bead-blasted by Lin-Dec, of St Ives, Cambs.

Although the Sebring motor looks absolutely straight-forward, many special tools are required to do the job properly. To save a lot of heartache, and possibly engine damage, anyone lucky enough to come into the possession of one of these little singles (there's a whole range of models, some with desmo heads) should send the motor away to be rebuilt properly rather than attempt the job at home. Engine parts can be expensive, though, so it's always a matter for serious thought.

Peter Kelly

Engine

Lifting the cylinder head from the Ducati was a piece of cake; but then it turned out to be the only job that was! To break open the top half of the little single you simply expose the timing gears, line up the dots and unbolt the head. There are only four bolts to remove and the head and barrel can be lifted off.

On our engine we knew that we were in for trouble right away because the combustion chamber in the head was black and sooty, and more than just a little wet. The problem appeared to be oil leaking past the piston rings.

When the oil was drained, we also found a lot of water in with it, a sure sign that the motor had been standing for some time without the benefit of any shelter from the weather. It was more like treacle than engine oil. Pulling off the first side cover involved the first special tool. The cover has an outrigger bearing and has to be pulled off dead square.

The clutch plates were all gummed up and had to be prised apart, but with a clean-up they could probably be re-used. Next the engine had to be locked up to remove the primary drive and clutch centre nuts. This involved the use of another special tool to lock things up. Without it you really have your work cut out.

After unlocking the nuts we found that the thread had stripped on the clutch nut but the thread on the shaft was okay. Apparently, the first tooth on the kickstart gear has a habit of breaking with this particular model but the one in our engine was all right — which was the first good news so far.

To strip the cylinder head, there are yet

This used to be a valve stem oil seal, if yours looks like this — replace it.

Yet another puller has to be used to shift the rocker arm spindles from the head.

more pullers to be used. The spindles have to be yanked out to remove the rocker arms. Also, after removing the bevel gear support bearing, it is well worth checking for signs of the outer race creeping — another common fault. Once the valves were out they were inspected and passed as fit for service but the springs would have to be renewed. On the Sebring and Monza models the valve springs are of a different rate for the inlet and exhaust so don't mix them up.

For a fairly simple single cylinder motor, there are so many special tools needed to strip the bottom half of the motor that tackling it on an adjustable spanner basis is just not on. However, one point that we did like was the shimming of the camshaft bevel gears. On each of the gear sets there is a good away section. To obtain the correct meshing of the gears you add or subtract shims until the ground areas are exactly smooth in relation to each other.

Apart from expected wear and tear on the motor's internals we did come across a crack in the off-side half of the crankcase. This had to be cleaned out and packed off to a specialist for welding, then machined back into shape — it worked out cheaper this way than buying a new pair of crankcases. If the damage had been more extensive we would have had to raid Mick Walker's "cave" for the relevant bits!

There will be another article on the Ducati when the rebuild has been completed.

Dave Walker

This special puller MUST be used to extract the gearbox bearing from its blind hole.

The factory way of locking the clutch drum and crankshaft — yes, you really do need these tools.

It is also impossible to remove the bevel drive bearings without this puller.